An Introduction to Constraint-Based Temporal Reasoning

Synthesis Lectures on Artificial Intelligence and Machine Learning

Editor
Ronald J. Brachman, *Yahoo! Labs*
William W. Cohen, *Carnegie Mellon University*
Peter Stone, *University of Texas at Austin*

Representation Discovery using Harmonic Analysis
Sridhar Mahadevan
2008

Essentials of Game Theory: A Concise Multidisciplinary Introduction
Kevin Leyton-Brown and Yoav Shoham
2008

A Concise Introduction to Multiagent Systems and Distributed Artificial Intelligence
Nikos Vlassis
2007

Intelligent Autonomous Robotics: A Robot Soccer Case Study
Peter Stone
2007

An Introduction to Constraint-Based Temporal Reasoning

Roman Barták, Robert A. Morris, and K. Brent Venable

ISBN: 978-3-031-00439-1   paperback
ISBN: 978-3-031-01567-0   ebook

DOI 10.2200/978-3-031-01567-0

A Publication in the Springer series
SYNTHESIS LECTURES ON ARTIFICIAL INTELLIGENCE AND MACHINE LEARNING

Lecture #26
Series Editors: Ronald J. Brachman, *Yahoo! Labs*
                William W. Cohen, *Carnegie Mellon University*
                Peter Stone, *University of Texas at Austin*
Series ISSN
Synthesis Lectures on Artificial Intelligence and Machine Learning
Print 1939-4608   Electronic 1939-4616

An Introduction to Constraint-Based Temporal Reasoning

Roman Barták
Charles University in Prague

Robert A. Morris
NASA Ames Research Center

K. Brent Venable
Tulane University and Florida Institute of Human and Machine Cognition

SYNTHESIS LECTURES ON ARTIFICIAL INTELLIGENCE AND MACHINE LEARNING #26

ABSTRACT

Solving challenging computational problems involving time has been a critical component in the development of artificial intelligence systems almost since the inception of the field. This book provides a concise introduction to the core computational elements of temporal reasoning for use in AI systems for planning and scheduling, as well as systems that extract temporal information from data. It presents a survey of temporal frameworks based on constraints, both qualitative and quantitative, as well as of major temporal consistency techniques. The book also introduces the reader to more recent extensions to the core model that allow AI systems to explicitly represent temporal preferences and temporal uncertainty.

This book is intended for students and researchers interested in constraint-based temporal reasoning. It provides a self-contained guide to the different representations of time, as well as examples of recent applications of time in AI systems.

KEYWORDS

temporal reasoning, constraints, scheduling, planning, uncertainty, preferences, applications of temporal reasoning

Contents

Preface

Physics teaches us that time is fundamental to the understanding of the material world. Time provides a substrate for the organization and observation of experience. In particular, time provides a way to order things that change the world, i.e., events and actions, as well as a means to measure the duration of those actions and events. These fundamental notions provide the basis for deductive logical systems for reasoning about duration and order. Since almost the dawn of computer technology a notion of time has been integrated into computer systems for database design, simulation, and Artificial Intelligence.

The development of temporal reasoning systems in AI has matured to the point that time is fully integrated into many industrial applications, primarily in planning and scheduling. It is fairly uncontroversial to state that these applications share a common "core" of accepted models and algorithms. The core models represent properties of order and duration in terms of constraints, which are typically organized in graphical form, where nodes are time elements (typically, but not always, representing the start or ends of events) and edges are labeled with temporal constraint information among the elements. The algorithms implement deductive temporal reasoning as operations on the constraint network, drawing upon previous results in graph theory to make navigation as efficient as possible. Years of productive academic research has refined and optimized these models and algorithms to the point that they satisfy the stringent performance requirements of industry and space.

At the same time, the field of temporal reasoning in AI continues to be active, partially in response to the complexity of the applications into which the temporal reasoning systems have been embedded. For example, autonomous robotic systems often execute actions in an uncertain and partially unknown environment. In such a world, the duration or ordering of events, including their own actions, may only be approximated prior to their being executed. This challenges the assumption of the traditional core theory that events and orders are deterministic in the world, and a rich body of recent work has focused on extending the core models and algorithms to handle this kind of uncertainty.

Another requirement of temporal reasoning systems that has evolved as the result of "lessons learned" from industrial applications is the need to view problems such as planning as optimization problems. In many applications, candidate plans have an implicit ordering with respect to criteria involving time: for example, although two plans or schedules might both be feasible, one might be better than another because one has a smaller makespan, or because one plan allows more time for some desired activity, such as science observations on a robotic space explorer. This requirement again challenges assumptions made by the core temporal model: in this case the assumption is that the constraints defined or derived by the problem are on equal footing with respect to their adherence by the temporal reasoner. Again, there is currently an active community at work expanding the core framework by imposing a "preference superstructure," which enables the ordering of temporal constraints, and thereby solutions, in terms of a set of desired criteria.

This book is intended to be a technical introduction to the representation of time typically found in AI systems. The remaining chapters are organized as follows:

1. Chapter 1 describes the mathematical and logical foundations of formal concepts of time and the introduction of automated systems for planning and scheduling.

2. Chapter 2 introduces the core components of constraint-based temporal reasoning systems.

3. Chapter 3 summarizes extensions to the core models to include reasoning about uncertainty and preferences.

4. Chapter 4 contains examples of applications of temporal reasoning in planning, plan execution, natural language processing and interpreting data.

This book is intended to be of interest to anyone wanting to grasp the fundamentals on constraint-based temporal reasoning. The book will be completely self-contained and will be made accessible to anyone with a minimal scientific background. In addition to being of interest to undergraduates, graduate students, as well as established researchers in computer science, we foresee it as being particularly useful for the large community of industry specialists working on planning- or scheduling-related real life problems.

Roman Barták, Robert A. Morris, and K. Brent Venable
December 2013

Acronyms of Temporal Reasoning Systems Discussed in this Book

The following table lists the acronyms identifying temporal reasoning systems, what they stand for, and the section in the book in which they are introduced.

Acronym	Meaning	Secction
PA	Point Algebra	2.1.1
IA	Interval Algebra	2.1.2
STP(N)	Simple Temporal Problem (Network)	2.2.1
TCSP	Temporal Constraint Satisfaction Problem	2.2.2
DTP	Disjunctive Temporal Problem	2.2.3
(S)TNA	(Simple) Temporal Network with Alternatives	2.2.4
QA	Qualitative Algebra	2.2.4
STPP	Simple Temporal Problem with Preferences	3.1.3
STPU	Simple Temporal Problem with Uncertainty	3.2.1
CTP	Conditional Temporal Problem	3.2.3
CSTNU	Conditional Simple Temporal Network with Uncertainty	3.2.4
STPPU	Simple Temporal Problem with Preferences and Uncertainty	3.3.1
CTPP	Conditional Temporal Problem with Preferences	3.3.2

CHAPTER 1

Introduction to Time in AI Systems

Physics teaches us that time is fundamental to the understanding of the material world. Continuous physical processes like waves, flows, and oscillations are characterized in terms of changes, and rates of change, of the world over time.

Time provides a mental substrate for the human management of perception and action. In particular, time provides a necessary cognitive and linguistic component for describing change. Change happens through the occurrence of events, processes, and actions, and time provides a way to record, order, and measure the duration of these occurrences. Indeed, periodic events and motion have long served as standards for units of time. Currently, the international unit of time, the second, is defined in terms of radiation emitted by caesium atoms. [118]

Time *management* is a fundamental aspect of intelligent behavior. Time management consists of describing, predicting, and planning actions or events. Artifacts for time management such as sundials, mechanical clocks, and calendars, arose from the need to record and predict when events or natural processes recur. The human organization and visualization of time in linear form, as points on a line, has ancient origins [96]. But it was primarily the modern industrial age that gave rise to the human awareness of what makes time management mentally *complex*. Computing machines arose as a natural technology to apply to solve problems in time management. In this chapter, we briefly trace the development of time management problems, their formulation in mathematics and logic, and the rise of algorithmic solutions, including those based in AI. We conclude this chapter with an overview of the goals and scope of this book.

1.1 THE RISE OF TIME MANAGEMENT: PLANNING AND SCHEDULING

Historically, it appears that the first hard problems in time management were planning and scheduling. The hardness of these problems arises from the interaction of three components:

- a set of *goals* (e.g., goods or services that need to be provided by a certain time), each requiring a sequence of tasks that must be performed in a certain order;

- a set of *resources* that are allocated to performing the tasks with constraints dealing with capacity or load (e.g., machines, people); and

- an *operating cost* (e.g., money or time itself) that cannot exceed a certain value and should be optimized (minimized or maximized) in the execution of the tasks.

Complex planning and scheduling have many specializations. For example, if the goal is a creation of a set of goods, the problem is *production* planning and scheduling; if the goal is to provide a transportation service, then the problem is *transportation* planning and scheduling. Scheduling for transportation includes timetabling, vehicle assignment, and crew scheduling, [49] and is associated with complex constraints, such as *only 20% of crew duties must be longer than 9 hours*, as well as optimization criteria such as *minimizing total idle time for the vehicle fleet*. A different timetabling problem is also defined in education, consisting of assigning classes (tasks) to instructors (resources).

Frederick Taylor was perhaps the first person to separate the problems related to planning from those of plan execution in manufacturing operations. This distinction enabled the rise of the scientific study of planning (in the broad sense that includes scheduling). Around the time of the First World War, he proposed the formulation of production planning offices for creating plans, monitoring inventory, and monitoring operations. Henry Gantt developed his famous bar charts around the same time as an aid for production control. These charts associated time (on the horizontal axis) with different quantities to be measured, such as tasks in a job, or the amount of work performed by human or machine [19]. Gantt was aware of the need to coordinate planning to avoid what he called "interferences," what we today in AI might call "constraint violations."

Other tools for production planning included loading, planning boards, and lines of balance [45]. Loading is a scheduling technique that computes the amount of capacity of resources required over time; loading can be infinite or finite, depending on whether the available capacity to perform the tasks is considered. Planning boards (also called control or schedule boards) are visual means for representing machine utilization or plans. Finding effective visualization tools for time management pre-dated the rise of computers, and continues as a separate design problem to this day. Although beyond the scope of this book, effective human-planning tool interaction is an important factor in the successful deployment of automated planning systems. The field of *mixed initiative planning* [15] arose in order to design systems that effectively combine human and machine decisions for temporal planning.

1.2 LOGICAL AND MATHEMATICAL FORMULATIONS

The survey of representations and algorithmic methods for solving time management problems contained in this book arose from concepts developed from years of research in mathematics, logic, operations research, and computer science. In this section we sketch a partial history of formulating problems of time management, starting in the 1940s with mathematical formulations of scheduling, through the use of temporal logics for representing change, and ending with the first set of approaches in Artificial Intelligence. The reasoning systems surveyed in later chapters of this book are algorithmic refinements of the methods surveyed in this section.

1.2.1   MATHEMATICAL FORMULATIONS

Automated production scheduling systems began in the mid 1950s out of the need to assist human schedulers in managing the "critical path," (the longest necessary path through a network of activities ordered in time) and perhaps equally, for the need for computer companies to find new useful things to do with their products. The Critical Path Method (CPM) and PERT used linear programming to represent time and the constraints between events in production sequences. A CPM schedule was visualized as a graph with nodes and edges labeled with temporal information. Algorithms such as Johnson's rule for the two-machine flowshop, earliest due date (EDD) rule for minimizing maximum lateness, and the Shortest Processing Time (SPT) rules for minimizing average flow time, were developed during this period.

Network flow and integer programming arose in the 1960s to solve problems involving time. Network flow problems are mathematical formulations of a wide range of practical problems involving the movement of some entity [49]. In project management, and other planning and scheduling problems, the entity of interest is time. In the *minimum project duration* flow problem, a set of jobs with duration and precedence relations are provided, and the problem is to find a set of start times for the jobs that minimize the overall project duration. This can be mapped into a shortest path problem in acyclic graphs. Second, *just-in-time* scheduling introduces additional constraints that require start times of some jobs to happen within a certain time after the start times of others. Third, problems with the *time-cost* tradeoff allow for the expression of time-cost tradeoff curves, which measure the increase in cost (e.g., manpower) that results from "crashing" a job (finishing it in the shortest possible duration). Solving this problem requires the assignment of optimal durations to the jobs for minimal cost, given the other temporal constraints.

Another way of introducing time into a flow problem is to transform static network models into *dynamic models* using *time-expanded networks*. The result is a class of *dynamic flow* problems. For example, in a static flow problem an edge between a pair of nodes in a network might represent a capacity of some sort, and the associated dynamic flow problem might be to determine the maximum total (transient) flow that can be sent between the nodes over a given period. To solve dynamic problems, a *time-extended replica* of the associated graph must be constructed, where p copies of the nodes are introduced, where p is the time period of interest. This technique is similar to the one employed in dynamic Bayes networks, as we'll see.

1.2.2   LOGICAL AND PHILOSOPHICAL FRAMEWORKS

The other main formal basis for the approaches discussed in this book is logic. The starting point here is the integration of time and truth in deductive systems, which began in the modern age in the 1950s with Arthur Prior [94]. Based on interpretations of ancient philosophers, Prior noticed that time can be treated as a *modality*, something that qualifies the truth of a proposition. Axioms and rules of temporal inference can be formulated from four tense operators, representing what has been true (always or at some instant) in the past or future. These axioms and rules extend a formal language, like the propositional or predicate calculus. One extension to basic tense logic of interest here is *metric* tense logic. This allows for deductive reasoning about durations, such as *It happened n time units ago that P.*

The semantics of tense logic follows closely the structure of other modal logics. A *temporal frame* $\langle T, < \rangle$ is a set of times and an ordering $<$ representing the "flow" of time. T may be the real numbers, if time is viewed as continuous, or the integers, if a discrete view of time is preferred. The tense logic axioms and inference rules can now be interpreted as expressing universal properties of the flow of time. An interpretation of a set of formulas assigns truth to each formula in all times in T, and the theorems of the logic are precisely all formulas that hold in all interpretations over all temporal frames. In addition to linear models of time, other formalisms model the uncertainty of the future through a notion of branching time.

An alternative to tense logic that has been influential in AI planning systems is the *method of temporal arguments*. Here, an expression in first order logic is extended to include a temporal argument (e.g., walked-to(John, market, 1PM)). If a predicate for $<$ and a constant "now" are added, then the axioms of tense logic can be "simulated" using first-order logic [42].

1.2.3   ORIGINS OF TIME IN AI SYSTEMS

The computational model of time in Artificial Intelligence recognizes the time-varying nature of much of human knowledge, as formalized in tense logics. Additionally, the impetus for early AI approaches to production scheduling included the failure of traditional OR optimization methods to scale well to large problems, as noted by Simon [106]. The re-formulation of computational problems as problems involving search through a space of decisions in AI allows the potential for explicitly managing search decisions through heuristics, which held out hope that these methods would allow for better scalability.

Representing change, especially change associated with actions, led to the *Situation Calculus*, an extension of first-order logic that allowed for the expression of axioms about actions and results of actions. Yet the Situation Calculus has no explicit representation of time; rather change was represented via the quantification over states or situations, complete descriptions of slices of the world. Because states are complete descriptions, one also required a complete description of the effects of actions, which in turn led to the infamous Frame Problem. In addition, it was awkward, if not impossible, to represent in the situation calculus actions with durations, or actions that overlapped with each other.

The *Event Calculus* [60] is based primarily on a simple change to the situation calculus, namely, replacing situations with time points as the objects of quantification. Fluents are true at a point if something like an action made the fluent true in the past and nothing has happened since to "clip" the fluent in between. An explicit representation of time also allows for a representation of processes ("liquid events") like walking, as well as discrete events. An explicit representation of time also allows for a representation of time durations as pairs of time points, and of systems for dating events.

Although tense logics have been useful in applications in natural language, and in the formal semantics of programs, they have not been found to be useful in other areas of AI. The main reason seems to be that they do not provide a time granularity adequate to capture the change in the world required for AI applications [3].

1.3 TIME GRANULARITY

Earlier we defined a simple mathematical representation of time as a pair $\langle T, < \rangle$ comprised of a set of time points T and an precedence ordering. T can either be dense (mapping to the reals) or discrete (mapping to the integers). This simple formulation of time is used implicitly in the temporal reasoning systems described in the next two chapters. By contrast, in human discourse and reasoning about time, clocks and calendars are employed, which impose levels of granularity on time. Using the terminology in [10], clock or calendar elements (seconds, minutes, days, business-months, etc.) are examples of *temporal types*, formally defined as a mapping between elements of an index set I into a collection of elements from an *absolute time set A*. For example, if A is a set of real numbers representing the time units of the year 1952, then **month** is a temporal type, and **month**(1) can be defined as the subset of A making up January 1952. Alternatively, using an example from Chapter 2, if A is the set $\{1, \ldots, 1440\}$ of minutes in a given day, then **hour**(6)**minute**(40) is the minute in A corresponding to 6:40AM, i.e., minute number 240.

Different temporal types are related hierarchically: for example, *business month* is a subtype of *month*, and *days* is finer-grained than *week*. Humans also express temporal constraints using temporal types: for example, we might be told that an online purchased item will be received within 10 business days from today. Much of the reasoning we perform using temporal types consists of converting between different temporal types: for example, we might be interested in inferring the total number of days it will take the item to be shipped, given that it will arrive 10 business days from today.

Temporal reasoning systems based on constraints, the focus of the next two chapters, have been extended to define reasoning problems with constraints over heterogeneous temporal types (for example, [10], [39]). Levels of granularity provide humans with structure that makes discourse and reasoning about time more efficient, and automating them offers challenging computational issues: therefore, it is important to investigate the formal theory of temporal types and reasoning with them. However, these issues are beyond the scope of this book. All the systems defined in what follows associate time with some notion of absolute time T with no additional temporal typing, and all constraints are expressed on the absolute time domain (indeed usually the systems are general enough that it holds over any specification of T). No data conversion is required. Roughly speaking, the view here is that a computer, unlike a human, is able to solve hard reasoning problems without the "syntactic sugar" of temporal types. It is these core computational problems that this book is interested in surveying.

1.4   TIME AND AGENT ARCHITECTURES

The modern view of AI is as the field that designs and engineers *agents*, where an agent is broadly defined as any system that perceives its environment through sensors and acts on the environment through actuators. An agent is also associated with a *performance measure* and a concept of *rationality*, defined as the ability to always exhibit behavior that is expected to maximize its performance measure [98]. An agent contains a *function* that maps a current percept (and other knowledge) into an action (see Figure 1.1). An agent acts in an environment, which can be viewed

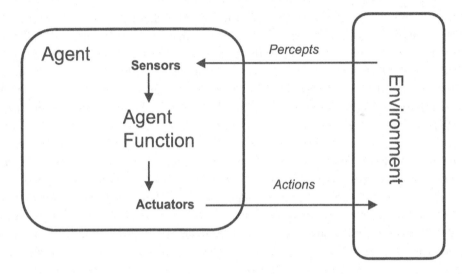

Figure 1.1: Simple Agent Architecture. From [98] .

abstractly as a description of the problem to be solved by the agent. For example, the task environment of a self-driving taxi is a grid of streets joined by intersections controlled by stop lights, navigated by cars, as well as houses and other landmarks of interest.

To appreciate the role of time in agent behavior, we refine the agent architecture. Figure 1.2 expands the agent concept along the time dimension for illustration. The agent continuously perceives and acts on an environment that changes over time. Agent activity is encapsulated in a *control loop* comprised of sensing the world, updating its world model and plan, and executing some action. We use the term *temporal abstraction* to refer broadly to an interpretation task: given a set of time-stamped data, produce abstractions (patterns) of the data that are relevant to agent goals [103]. A temporal abstraction could be as simple as a robot recording a simple event using an Allen relation (*hitting a wall followed moving forward*) or as complex as recognizing a recurring pattern (*For patient P, stomach ache always followed ingesting medication M after 2 hours*). Abstraction is used for prediction and for organizing sensing activity (e.g., monitoring or tracking a certain state or fluent), as well as for updating plans.

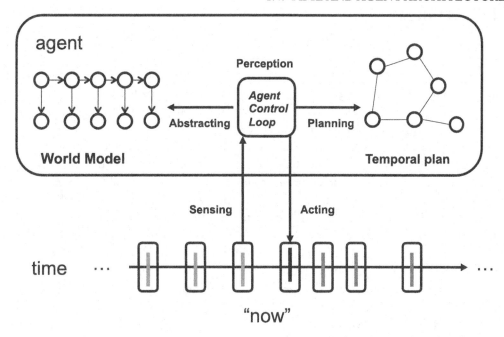

Figure 1.2: An agent continuously perceives, modifies its world model, and acts in time.

In addition to temporal planning and abstraction, we introduce a notion of *temporal perception*, the awareness of the passage, or flow, of time. It has been noted that the feeling of time dramatically affects the dynamics of human cognition [71]. A simple example of the need for time perception is that the mere passage of time can enable an action in an agent plan to become executable. Without some notion of time perception the agent would not know this. More broadly, the perception of time in artificial agents improves agent robustness (i.e., resilience from failure) by imposing a sort of "meta-control" of the agent's control loop. Relevant examples include execution latency [81], the gap between when an action is enabled and when it's executed by the agent; negotiation under time constraints by distributed agents [37]; and anytime algorithms. Roughly, the rational use of time perception allows an agent to allocate the appropriate amount of time to each element (sense-decide-act) of the inner control loop, depending on the current response-time requirements.

Therefore, agent representations of time are needed

- to abstract temporal information as part of monitoring and prediction;

- to formulate and update its activity plan; and

- to maintain execution robustness in the presence of the passage of time.

1.4.1   EXAMPLE

We introduce a simple self-driving taxi example to illustrate the role of time in agent design. Consider first a very specialized driving capability: *applying the brakes when the vehicle immediately in front applies its breaks*. Although specialized, the performance metric is complex. It's not a simple reflex action of applying the break immediately after seeing the break lights of the car in front (we can ignore the anomalous case and assume the break lights always work). Skillful drivers consider at least two questions: how fast do I need to apply the breaks, and how hard. The answers of course depend on the perception of distance, and perhaps on the perception of how hard the driver in front is applying his brakes (e.g., do you see smoke from the tires or swerving from the car in front?). Already the driver seems to be implicitly reasoning about the *time to collision* from the onset of brake action, and by inserting either a brake action immediately, or perhaps a sequence comprised of a "watchful wait" action to see if the distance between the cars gets critical in the near future, followed by a brake action. The driver could be also reasoning about the "dose" of brake to apply: he wants to avoid multiple sudden jerky actions if possible, and would prefer a single dose over as long a period as necessary to resume a safe state.

This simple example shows that time enters potentially in three places in the agent sense-decide-act loop: in the sense phase, when the agent measures the rate of change of distance over time; in the decide phase when it reasons about how much time it has to respond; and in the action phase, when it either inserts a reflect brake action or a wait-followed-by brake sequence. In general, the agent must have a *model* of how distance shrinks with time after a braking action by the forward car, and use this model to guide the braking response.

If we focus merely on the autonomous braking sub-system of this example, we might miss what makes time critical in complex agents like taxi drivers. In this example so far, physics tells us that time and distance are related simply through velocity and acceleration, so the agent capability can be engineered without the agent explicitly having a "time-based model." For example, in the PID controller, each of the three gain parameters have a temporal interpretation as current error (P), duration of accumulated error (I) and predicted (future) error (D). Tuning these parameters properly results in the desired behavior of the controller without having explicitly setting or predicting values for time during control loop execution.

Where time enters more crucially is when we design complex systems that are collections or organizations of sub-models, such as the full taxi agent. First, a taxi agent is *goal-directed*: it allows for a representation of delivery goals *Take X from A to B* and has the means to automatically decompose goals into sequences of actions (plans) to accomplish them. The performance measure for such a goal-directed system might explicitly mention time (e.g., shortest time path on a navigation system). When the environment includes traffic, time and distance are more indirectly related, because there are other processes (e.g., traffic flow) that are involved. Unlike the breaking example, to accomplish goal-directed behaviors where time is a performance metric, it is generally required to have an explicit representation of time *as a universal measure of duration and order.*[1]

Second, a complex agent is usually *utility-based*, in which the performance metric involves a notion of utility or "happiness." In the braking example, the procedure for performing the responsive action may depend on whether the cargo is human or freight: the performance metric may depend on the comfort of the cargo, which is generally not involved for freight. The number, time of onset, and duration of braking actions may be the considering factors in measuring comfort.

Consider another, more difficult example, an emergency medical transportation operations agent. Here the driver (or team) transporting a patient to a hospital must have a model of two very different processes: traffic flow and patient condition. Events and actions that affect the former are based on different physical processes than the latter, and yet they must be combined in order to ensure rationality on the part of the medical team. The single quantity that binds events of all types together is time; it's virtually the only variable that can be used to relate the actions needed to accomplish the goal of safe transport. Again, time is a "universal category."

To summarize this introduction:

- In the agent view of AI, time enters into modeling the rational performance of agents in potentially three ways: in ordering potentially complex collection of percepts using a temporal model of abstraction; in recognizing the response-time requirements of the percepts (perceiving the flow of time); and to choose action (sequences) to respond to the percepts, possibly applying a temporal plan.

- Time becomes indispensable in complex, goal-directed, or utility-directed agents in which the performance metric is a complex combination of criteria that are based on different natural processes. Time, through a characterization of temporal duration, order, or flow, is a means for ordering or measuring change in the state of the world.

1.4.2 OVERVIEW OF REMAINDER OF BOOK

The objective of this book is to provide a concise but comprehensive introduction to the major computational approaches to time in artificial intelligence systems. For a more detailed introduction to many of these systems, the reader is encouraged to consult [38].

Chapter 2 reviews the "core" models of time and temporal reasoning. These core models can be distinguished based on whether they offer a *qualitative* or *quantitative* approach to time, and also whether the temporal relations are between *instants* of time or *intervals*. Hybrid models combining qualitative-quantitative or point-interval reasoning, are also covered.

Chapter 3 explores two major extensions to the core models: through the introduction of *uncertainty* reasoning, or through the introduction of reasoning about temporal *preferences*. Uncertainty reasoning is important, for example, in making predictions about an unknown future;

---

[1]This is perhaps what Kant and Aristotle had in mind when they characterized time as a *universal category.*

a practical example is an automated trading agent for answering questions such as *During the next two months, when is the best time to buy a certain stock?* [28].

A temporal reasoning problem based on constraints may admit of no, one, or many solutions. If there are no solutions (in which case the problem is said to be *overconstrained*) then we may be interested in finding ways of relaxing the constraints so that a solution can be found. Alternatively, if there are many solutions (i.e., the problem is *underconstrained*) then we may be interested in finding ways to order the solutions based on some criteria of *quality*. To automate either of these decisions (what constraints to relax or how to order the solutions), the temporal constraint model needs to be modified in order to express *preferences* with respect to solutions or constraints. These extensions are also explored in Chapter 3.

Finally, in Chapter 4, we explore examples of deployed systems that utilize one or more of the temporal reasoning models and languages explored in Chapters 2 and 3. We look at applications in the medical service industry and other industrial applications, as well as applications in space. The applications range from capabilities in temporal perception, temporal abstraction, as well as planning and scheduling, as described in this chapter.

CHAPTER 2

Temporal Frameworks Based on Constraints

The representation of temporal information, and reasoning about time, are important in artificial intelligence. Reasoning about time plays an important role in building automated planning and scheduling systems, where causal and temporal relations are the most critical concepts. Time also enters into areas such as natural language processing, for example to represent stories. In these areas we need to express information about events and processes such as "reading newspapers" and "having a breakfast" that span over an interval of time and that are related to each other, as expressed in sentences such as "I read the newspaper during breakfast."

A time-aware rational agent abstracts information about past events and observed processes from the sensors. These *temporal references* (time points and intervals) and *temporal relations* between them (such as "some event happened before another event" and "a given process had a specific duration") are stored in a *temporal knowledge base*. To do reasoning such as activity planning, the agent adds temporal references about future events (goals, etc.) and futures processes (planned activities) and asks if all these temporal references and relations are consistent. This way the agent infers what may happen in future, what is inevitable, and what is impossible.

More formally, any system that contains an explicit representation of time contains *temporal references* (time points and intervals) and *temporal propositions* describing the temporal relations between the temporal references. These components combine into a *temporal reasoning system* consisting of

- a *temporal knowledge base* containing temporal propositions,

- a procedure for *checking consistency* of the temporal propositions, and

- an *inference mechanism* that is used to deduce new information and answer queries about the temporal references.

In this chapter we will describe the core frameworks for representing temporal references and for expressing temporal propositions. We will also introduce the basic algorithms for consistency checking and we will survey the complexity results. We will abstract from the mechanisms to obtain temporal information (from sensors, etc.) and to use it in further reasoning (such as activity planning). The focus will be on formal models of time and on temporal reasoning in these models.

There exist two core approaches for temporal reasoning:

- a *qualitative* approach focuses on relative temporal relations, for example some event happens before another event,

- a *quantitative* approach focuses on metric (numerical) temporal relations, for example some event happens at least two hours after another event and lasts for one hour.

We shall now describe both approaches in more detail.

2.1 QUALITATIVE TEMPORAL FRAMEWORKS

The qualitative approach to the representation of time is based on relative temporal relations between the temporal references. Assume the following situation:

> *"I read newspapers during breakfast and after breakfast I walked to my office."*

Notice that there is no numerical (quantitative) information included in the description of the situation. Nevertheless we can still do some reasoning and deduce useful additional information that is not explicitly included in the description. In particular, we can represent this situation using two types of temporal references:

- *temporal intervals* representing the activities "reading newspapers", "having a breakfast", and "walking to office"; and

- *time points* representing the important events, in this case the starts and ends of the activities.

These temporal references induce a connection using temporal constraints that describe relative positions of the temporal references in time. Figure 2.1 introduces a timeline on which the activities from the above example are placed in their relative positions. It also shows two *constraint networks* representing the situation: one with the nodes of the network standing for the temporal intervals (a) and one with the nodes standing for start and end points of the events (b).

Having a representation of available knowledge, we are typically interested in determining whether the given information is *consistent*, that is, whether it is possible to assign exact times to time points without violating the constraints. Such a consistency check can then be used to deduce additional information about the situation. Thus, we can ask *"Do I read the newspapers while entering the office?"*. Answering this question can be realized by adding a new temporal constraint representing the question to the constraint network (the reading activity finishes after the end of the walking activity) and validating the consistency of this new network. If the network is consistent then the situation is indeed possible, although it does not mean that the situation is inevitable, i.e., that the constraint is entailed by the network. To ensure that some constraint is inevitable, for example if the reading activity must finish strictly before the walking activity, we can, as we will see later, find the so-called *minimal constraint network* where no redundant constraints are present. If the constraint is present in the minimal network, then it is indeed inevitable.

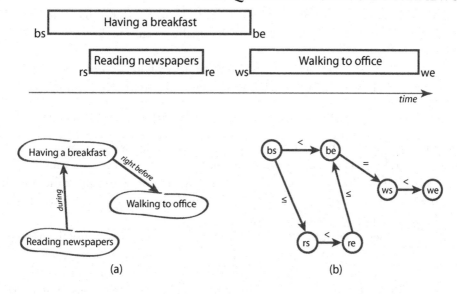

Figure 2.1: Example of several related activities and their representation using the constraint networks over the temporal intervals (a) and the time points (b).

In the next sections, we will present the formal frameworks for qualitative reasoning on temporal intervals and time points and formally define notions such as temporal constraint networks, consistency, and minimal networks. Two major qualitative temporal frameworks will be presented:

- the point algebra for working with time instants (points); and

- the interval algebra for working with temporal intervals.

2.1.1   POINT ALGEBRA

The point algebra (PA) is a symbolic calculus for working with qualitative ordering constraints between the pairs of time instants (time points). Since a time point represents an exact single event in time, there are only three possible ordering relations between a pair of time points. Assume that t_1 and t_2 are two time points then the following primitive relations are possible:

- $[t_1 < t_2]$, the point t_1 is before the point t_2;

- $[t_1 = t_2]$, both points t_1 and t_2 are identical; and

- $[t_1 > t_2]$, the point t_1 is after the point t_2.

Sometimes, the exact relative position of points is not known precisely. For example, we may only know that t_1 is not identical to t_2. Such relations can be described using sets of primitive relations;

for example the relation $[t_1 \{<, >\} t_1]$, abbreviated to $[t_1 \neq t_2]$, states that t_1 is either before or after t_2. Notice that the set of primitive relations actually means the disjunction of the primitive relations.

Let $P = \{<, =, >\}$ be a set of primitive relation symbols. To describe every relation between any pair of points we can use the following eight quantitative constraints:

$$R = 2^P = \{\{\}, \{<\}, \{=\}, \{>\}, \{<, =\}, \{>, =\}, \{<, >\}, \{<, =, >\}\}$$

The meaning of $\{\}$ is that the relation between the time points is inconsistent, i.e., there is no assignment of a relation between the two time points that is consistent with the available knowledge. The *universal relation* $P(= \{<, =, >\})$ means that no information about the current relation between the two points is known. To simplify the notation we shall use $\neq$ instead of $\{<, >\}$, $\leq$ instead of $\{<, =\}$, and $\geq$ instead of $\{>, =\}$.

As the relations are expressed as sets of primitive relations, we can use the usual operations on sets: $\cup$ (union) and $\cap$ (intersection) for inference. These set operations represent disjunction and conjunction of constraints on a pair of time points. Assume that constraints $[t_1 \ r \ t_2]$ and $[t_1 \ s \ t_2]$ hold. Then we can combine the information using a single constraint $[t_1 \ (r \cap s) \ t_2]$. Similarly, if we know that either $[t_1 \ r \ t_2]$ or $[t_1 \ s \ t_2]$ holds then we can use a single constraint $[t_1 \ (r \cup s) \ t_2]$. The third important operation on temporal relations is a *composition operation* that handles transitivity. We denote this operation $\circ$ and its meaning is as follows. Assume that $[t_1 \ r \ t_2]$

$\circ$	$<$	$=$	$>$
$<$	$<$	$<$	P
$=$	$<$	$=$	$>$
$>$	P	$>$	$>$

Figure 2.2: Composition operation for the primitive operations in the point algebra.

and $[t_2 \ s \ t_3]$ hold. Then the relation between the time points t_1 and t_3 is defined as $[t_1 \ (r \circ s) \ t_3]$. For example $[t_1 \leq t_2]$ and $[t_2 < t_3]$ implies $[t_1 < t_3]$. The composition table for all pairs of primitive operations is given by Figure 2.2.

The composition relation between non-primitive relations is defined as a union of compositions of all pairs of primitive operations from these two sets:

$$r \circ s = \bigcup_{p \in r, q \in s} p \circ q.$$

For example $[t_1 \leq t_2]$ and $[t_2 < t_3]$ gives $[t_1 < t_3]$ because $(\{<, =\} \circ \{<\}) = \{<\}$. Note that this definition of the composition operation has the distributivity property:

$$(r \cup q) \circ s = (r \circ s) \cup (q \circ s) \text{ and } s \circ (r \cup q) = (s \circ r) \cup (s \circ q).$$

Typically the combination of ∘ and ∩ operations is used in the following way. Assume that $[t_1\ r\ t_2]$ and $[t_1\ s\ t_3]$ and $[t_3\ q\ t_2]$ hold. This means that we know some explicit information about the temporal relation between t_1 and t_2 (r) and some implicit information about this relation given by the constraints s and q. We can naturally combine these constraints in the following way:

$$[t_1\ r\ t_2] \wedge [t_1\ s\ t_3] \wedge [t_3\ q\ t_2] \Rightarrow [t_1\ (r \cap (s \circ q))\ t_2].$$

This process can be naturally extended to any sequence of time points connected by the constraints.

Let us return to the example from Figure 2.1 (b). There are six time points $\{bs, be, rs, re, ws, we\}$ and six explicit temporal relations between these points representing the situation in the point algebra. We asked the question *"Did I still read the newspapers when entering the office?"*. This question can be reformulated as *"Did I start reading before finishing the walk and stop reading after finishing the walk?"* which can be translated to the constraints:

$$[rs < we] \wedge [we < re].$$

Using the composition operation applied to the path (re, be, ws, we) we can infer the constraint $[re < we]$. Note that if we have a constraint $[t_1\ r\ t_2]$ we can naturally formulate the constraint in the inverse direction $[t_2\ r'\ t_1]$. In particular, if $r = (<)$ then $r' = (>)$ and vice versa. If $r = (=)$ then $r' = r = (=)$. Hence, $[re < we]$ is identical to $[we > re]$. As both the inferred constraint $[we > re]$ and the constraint $[we < re]$ from the query must hold together (conjunction), we can intersect them which gives an empty constraint (the intersection of sets of primitive relations is empty). An empty constraint indicates inconsistency which means that the constraint $[we < re]$ is inconsistent with the described situation and the answer to our question is *"No, I did not read the newspapers when entering the office."* Formally, this reasoning process can be described as

$$(r_{re,be} \circ r_{be,ws} \circ r_{ws,we}) \cap r_{re,we} = ((\{=, <\} \circ \{=\} \circ \{<\}) \cap \{>\}) = (\{<\} \cap \{>\}) = \{\}$$

We can generalize the above ad-hoc reasoning process to a general PA consistency problem.

Definition: [116] Assume that we have n time points $X = \{t_1, t_2, \ldots, t_n\}$ and a set C of binary temporal constraints between them. Each constraint is a relation from R. We define a binary constraint network for the point algebra, called a *PA network*, as a directed graph (X, C), where each arc from C is labeled by a constraint $r_{ij} \in R$ such that $[t_i\ r_{ij}\ t_j]$. We say that a vector of real numbers $(v_1, v_2, \ldots, v_n)$ is a *solution* to a PA network (X, C) if and only if the values $t_i = v_i$ satisfy all the constraints from C. We say that a PA network is *consistent* if a solution exists.

Figure 2.1 (b) gives an example of a PA network and $\{bs = 0, rs = 0, re = 5, be = 7, ws = 7, we = 9\}$ is a solution to this PA network so this network is consistent. Notice that the assignment of values to the time points means that any pair of time points (t_i, t_j) is related by a single primitive constraint $p_{ij} \in r_{ij}$. These primitive constraints characterize the relative positions of time points in the solution. For example, $[bs = rs]$ holds in our example solution. There exists a nice characterization when a PA network is consistent.

Proposition: A PA network (X, C) is consistent if and only if there is a primitive constraint $p_{ij} \in r_{ij}$ for each pair (t_i, t_j) of time points, such that every triple of primitive constraints satisfies $p_{ij} \in p_{ik} \circ p_{kj}$.

The consequence of the above proposition is that we can decide the consistency of a PA network by looking for the primitive constraints $p_{ij} \in r_{ij}$ meeting the consistency condition rather than by finding the instantiation of t_i variables. If we view the r_{ij} as finite domains from which we are seeking a single element satisfying the consistency condition, we can use techniques from constraint satisfaction to determine consistency.

Definition: We say that a primitive constraint $p_{ij} \in r_{ij}$ is *redundant* in a PA network if there does not exist any solution in which $[t_i \; r_{ij} \; t_j]$ holds. We say that a PA network is *minimal* if it has no redundant primitive constraints among its set of constraints.

We already introduced the technique to remove the redundant primitive constraints via a transitive closure of temporal relations:

$$[t_1 \; r \; t_2] \wedge [t_1 \; s \; t_3] \wedge [t_3 \; q \; t_2] \Rightarrow [t_1 \; (r \cap (s \circ q)) \; t_2].$$

This technique is known as *path consistency* in constraint satisfaction and it is applied to PA networks as follows:

```
for k = 1, ..., n do
    for i, j = 1, ..., n do
        r_ij ← r_ij ∩ (r_ik ○ r_kj)
    end
end
```

The time complexity of the above path consistency algorithm is $O(n^3)$. Obviously, if any constraint becomes empty during the path-consistency process then the PA network is not consistent. This means that being path-consistent is a necessary condition for a PA network to be consistent. The nice result from a computational standpoint is that this necessary condition is also a sufficient condition.

Proposition: [64] A PA network (X, C) is consistent if and only if the path consistency algorithm does not make any constraint empty.

Consistency and solution generation of PA networks can also be accomplished in $O(n^2)$ [113]. Though path consistency algorithms can be used to verify whether a given PA network is consistent, they do not necessarily provide a minimal network. Figure 2.3 shows an example of a PA network that is path consistent, but it is not minimal because the equality constraint between nodes s and e is redundant—there is no solution where these two points can be identical. To obtain a minimal network, we need what is called a 4-consistency algorithm [114], which determines consistency over subsets of 4 nodes rather than triples.

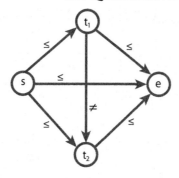

Figure 2.3: Example of a path-consistent PA network that is not minimal.

2.1.2   INTERVAL ALGEBRA

The second major qualitative approach uses intervals as temporal references and forms constraints from relative locations of pairs of intervals. Similarly to the point algebra, we can define an *interval algebra* (IA) that is a symbolic calculus for reasoning about relations between the intervals [2]. An interval i is defined by two real numbers (i^-, i^+), depicting their end points, where i^- is the start of the interval, i^+ is the end of the interval, and $i^- < i^+$. We can also see the interval as two points from the point algebra and the relative location of two intervals can be expressed using the constraints over the points. We will discuss the relation between the point algebra and the interval algebra later. Let us just mention now, that the interval algebra is actually an older concept [2] and the point algebra was proposed to cover some tractable sub-class of the interval algebra [116].

Let us consider two intervals i and j. There are seven basic relations that can hold between the pair of intervals: b(*efore*), m(*eets*), o(*verlaps*), s(*tarts*), d(*uring*), f(*inishes*), and e(*quals*) as depicted in Figure 2.4. Each of these relations has a symmetrical counterpart: for example if i is before j then j is after i. We will denote these symmetrical (inverse) relations by appending i to the names of the basic relations, bi for *after* (inverse to being before), mi for *is-met-by*, oi for *is-overlapped-by*, di for *includes*, fi for *is-finished-by*. The semantics of the inverse relations is fully given by the original relations where we swap the intervals. For example, $[i \ \{di\} \ j]$ means $[j \ \{d\} \ i]$. As the inverse relation to e(*qual*) is again e(*qual*), we have in total thirteen primitive relations $P = \{b, m, o, s, d, f, e, bi, mi, oi, si, di, fi\}$. All these relations can be described as relations between the two points representing the interval. For example the relation $[i \ \{d\} \ j]$ can be represented as $[j^- < i^-] \wedge [i^+ < j^+]$ (see Figure 2.4). Note also that these thirteen relations are all possible consistent cases that appear between the four points representing the two intervals.

If the relative location of two intervals is not known precisely, we can describe the relation as a disjunction of primitive relations. Similarly to the point algebra, we will use the set of primitive relations to represent this disjunction. Assume, for example, that we only know that intervals i

x **before** y	$x^+ < y^-$	
x **meets** y	$x^+ = y^-$	
x **overlaps** y	$x^- < y^- < x^+,\ \ x^+ < y^+$	
x **starts** y	$x^- = y^-,\ \ x^+ < y^+$	
x **during** y	$y^- < x^-,\ \ x^+ < y^+$	
x **finishes** y	$y^- < x^-,\ \ x^+ = y^+$	
x **equals** y	$x^- = y^-,\ \ x^+ = y^+$	

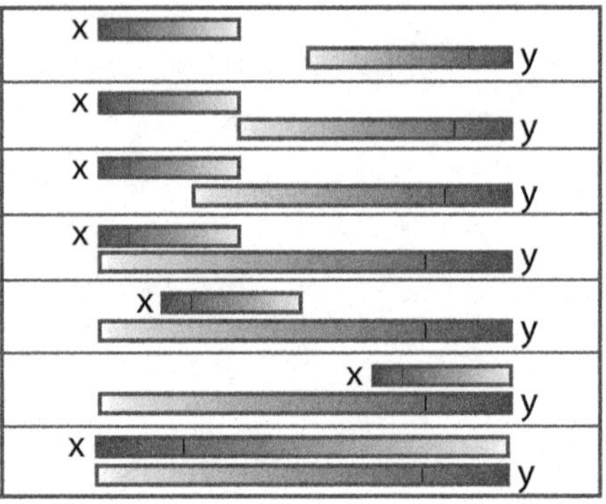

Figure 2.4: Primitive relations between two intervals.

and j are disjoint. This does not give us their precise relative location, but we can still represent this relation as [i {b, bi} j], either i is before j or i is after j (j is before i). There are in total 2^{13} possible relations between a pair of intervals:

$$R = 2^P = \{\{\}, \{b\}, \{m\}, \ldots, \{b, m\}, \{b, o\}, \ldots, \{b, m, o\}, \ldots P\}.$$

The empty relation {} represents an inconsistent situation, whereas P means that no information about the relative location of two intervals is known.

Let us recall the situation from our example and describe it accurately using the intervals and relations between the intervals:

"I read newspapers during breakfast and after breakfast I walked to my office."

There are three activities mentioned in the text, namely reading newspapers, having a breakfast, and walking to an office. These activities can be naturally represented as three intervals *Reading*, *Breakfast*, and *Walking*. The text also describes what the relations between the activities are. These relations can be described in the interval algebra as follows:

[*Reading* {s, d, f, e} *Breakfast*] $\wedge$ [*Breakfast* {m} *Walking*].

Notice that we translated the English word *during* using four primitive relations {s, d, f, e} because from the text, it is not fully clear whether I read all the time during the breakfast (*equals*) or just part of it (*starts, during, finishes*). Figure 2.5 shows the situation including all possible locations of the *Reading* activity.

Despite the fact that the precise times of the activities are not specified, we can still deduce some interesting information from the formal representation. This is very similar to the point algebra, but we are now working with intervals rather than with time points. For example, we can deduce that *Reading* does not overlap with *Walking*; more precisely *Reading* happens (right)

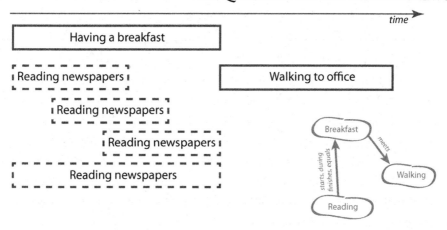

Figure 2.5: Example of temporal interval relations with a not fully specified relation.

before *Walking*. Formally [*Reading* $\{b, m\}$ *Walking*] holds and all other primitive relations between *Reading* and *Walking* are inconsistent. To infer this, we introduce a notion of *composition* of two relations similar to the one used in the point algebra. If S and T are two relations, such that [i S j] and [j T k] then we can compute the composition of relations $S \circ T$ that holds between the intervals i and k: [i ($S \circ T$) j]. Figure 2.6 shows the compositions of selected primitive relations. The composition of sets (disjunctions) of primitive operations is again done as a union of compositions of all pairs of primitive operations from these two sets:

$$ S \circ T = \bigcup_{p \in S, q \in T} p \circ q. $$

Notice that composition is not a commutative operation (i.e., the composition table is not symmetrical).

In addition to composition, we can again use the *intersection* of relations, representing the conjunction of relations, and the *union* of relations representing their disjunction. Both union and intersection are commutative operations. Again, intersection and composition are operations used for inference, i.e., they can be used to deduce new constraints via the the transitive closure:

$$ [i_1 \ S \ i_2] \wedge [i_1 \ T \ i_3] \wedge [i_3 \ Q \ i_2] \Rightarrow [i_1 \ (S \cap (T \ \circ \ Q)) \ i_2]. $$

Having represented the knowledge given as the input using the interval algebra framework, the typical reasoning task is determining whether the given information is consistent, i.e., whether it is possible to arrange the intervals along the timeline in such a way that all the relations hold.

Definition: Assume that we have n intervals $X = \{i_1, i_2, \ldots, i_n\}$ and a set C of binary temporal constraints between them. Each constraint is a relation from R. We define a binary

$\circ$	b	m	o	s	d	f	e
b	b	b	b	b	$\{b,m,o\}$	$\{b,m,o,s,d\}$	b
m	b	b	b	m	$\{o,s,d\}$	$\{o,s,d\}$	m
o	b	b	$\{b,m,o\}$	o	$\{o,s,d\}$	$\{o,s,d\}$	o
s	b	b	$\{b,m,o\}$	s	d	d	s
d	b	b	$\{b,m,o\}$	d	d	d	d
f	b	m	$\{o,s,d\}$	d	d	f	f
e	b	m	o	s	d	f	e

Figure 2.6: Composition operation for certain primitive operations in the interval algebra.

constraint network for the interval algebra, the *IA network*, as a directed graph (X, C), where each arc from C is labeled by a constraint $r_{ij} \in R$ such that $[i_i \; r_{ij} \; i_j]$. We say that a vector of pairs of real numbers $((i_1^-, i_1^+), (i_2^-, i_2^+), \ldots, (i_n^-, i_n^+))$ is a *solution* to a IA network (X, C) if and only if the intervals $i = (i^-, i^+)$ satisfy all the constraints from C. We say that an IA network is *consistent* if a solution exists.

Similarly to the point algebra, $\{Breakfast = (0, 7), Reading = (0, 5), Walking = (7, 9)\}$ is a solution to the problem from Figure 2.5. Again, each pair of intervals is related via a single primitive relation, for example $[Breakfast \; (s) \; Reading]$. All these primitive relations satisfy the property: $p_{i,j} \in p_{i,k} \circ p_{k,j}$.

Proposition: An IA network (X, C) is consistent if and only if it admits a *singleton labeling*, that is, if there is a primitive constraint $p_{i,j} \in r_{i,j}$ for each pair (i_i, i_j) of intervals, such that every triple of primitive constraints satisfies $p_{i,j} \in p_{i,k} \circ p_{k,j}$.

Similarly to the point algebra, we can define the *redundant* constraints and the *minimal* IA network. Path consistency can again be used to remove some redundant constraints and if any constraint becomes empty then the IA network is not consistent. Consequently path-consistency is a necessary condition for the IA network to be consistent. Unfortunately, path consistency is no longer a sufficient condition to ensure consistency of IA networks. Figure 2.7 gives an example of an IA network that is path consistent but it is not consistent.

As path consistency is not a complete technique to verify consistency of IA networks, to achieve completeness one is required to use standard constraint satisfaction techniques, for example backtracking combined with maintaining path consistency for this problem. This approach is viable as the consistency checking of IA networks is an NP-complete problem [116].

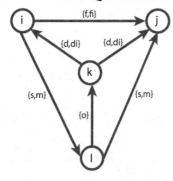

Figure 2.7: An example of inconsistent IA network that is path consistent.

2.1.3   TRACTABILITY OF SPECIFIC INTERVAL ALGEBRAS AND RELATION TO THE POINT ALGEBRA

There exists a nice geometrical interpretation of the interval algebra, where each interval is represented as a point in the Cartesian plane and the positions of these points correspond to temporal relations between the intervals. Assume an interval (i^-, i^+), with $i^- < i^+$. This interval naturally corresponds to a geometrical point $(i^-, i^+) \in \Re^2$. Conversely, each point $(x, y) \in \Re^2$ such that $x < y$ corresponds to an interval (x, y). Hence the subarea of 2D plane consisting of points $(x, y) \in \Re^2$ such that $x < y$ corresponds to all possible intervals and vice versa. Figure 2.8 shows this subarea with one highlighted point i.

An interesting property of this interpretation is that a temporal relation between two intervals can also be observed in the 2D interpretation. In particular, if we take a geometrical point representing the interval i, we can split the part of Cartesian plane where $x < y$ into six areas and six lines. Each area and each line represents all intervals that have a particular relation to the interval i. Figure 2.8 shows these areas and lines. For example the horizontal line containing the node representing i describes all intervals that finish at time i^+. If we take the part right from i then we obtain all intervals starting after i^- and hence this line represents the relation f. Symmetrically, the part of the line going to the left from i represents the intervals that start before i^- – they are in relation fi to i. The point corresponding to i itself represents the relation e. The remaining areas and lines can be identified with other primitive relations (see Figure 2.8).

Consider the constraint $\{b, m, o, s, fi\}$, and the area defined by all intervals (points) that are in this relation to interval i. This area forms a convex region in the Cartesian plane, that is, if we connect any two points in the region, then any point in the line between these two points also belongs to the region. The convexity property is important to identify sets of constraints such that the consistency can be decided in polynomial time. We can abstract from the Cartesian plane and define a graph G_{IA} to capture the convexity property of the constraints. This graph consists of thirteen vertices, where each vertex corresponds to one primitive relation in the interval algebra.

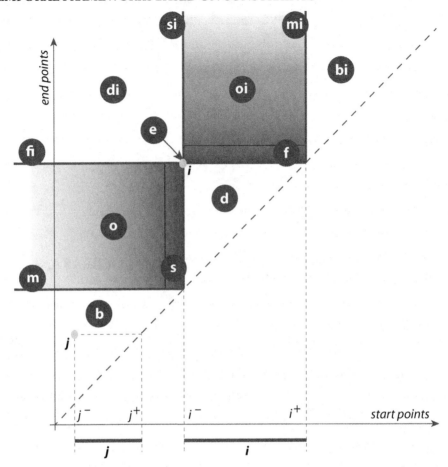

Figure 2.8: A geometric interpretation of interval algebra.

There is an edge between the nodes if and only if the areas in the Cartesian plane corresponding to the primitive relations are adjacent. For example, o is connected to m, s, and fi. Figure 2.9 shows the graph G_{IA}. We say that a constraint R (a set of primitive IA constraints) is *convex* if and only if for any two primitive constraints r and s from R, all constraints along the shortest path between r and s in G_{IA} are also in R. For example, the constraint $\{b, m, o, s, fi\}$ is convex, whereas $\{b, m, bi, mi\}$ is not. It is possible to show that the constraint R is convex if and only if for any interval i the area defined by all intervals j such that $j \, R \, i$ is convex in the Cartesian plane.

Out of 2^{13} constraints in the IA algebra, only 82 constraints are convex. If we restrict the interval algebra to these convex constraints we get an algebra called IA_c. This is indeed an

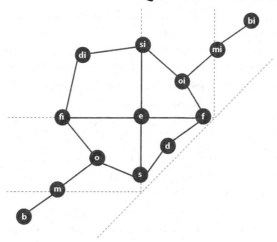

Figure 2.9: A G_{IA} graph representing the relations between intervals.

algebra because it is closed under the composition and intersection operations. In other words, if we compose or intersect two convex constraints then we obtain a convex constraint. It turns out that checking consistency of temporal network in IA_c is a tractable problem. Moreover, the path-consistency algorithm is complete and provides minimal networks in IA_c networks.

The notion of convexity can also be defined for the point algebra. We say that a relation R is *convex* in the point algebra if for any point t the set of points s such that $s\ R\ t$ defines a convex part in $\mathfrak{R}$. It is easy to show that the only non-convex constraint in PA is $\{<, >\}$. Similarly to IA we can define algebra PA_c over the convex constraints. Again, this algebra is closed under the composition and intersection operations. We already know that checking consistency of PA networks is a tractable problem. Moreover, for PA_c the path-consistency algorithms give minimal networks, that is, they remove all redundant primitive constraints.

There is an interesting relation between the IA_c and PA_c algebras. Each convex constraint from IA_c can be expressed as a conjunction of convex constraints in PA_c. For example the constraint $i\ \{s, e\}\ j$ can be expressed as: $i^- < i^+ \wedge j^- < j^+ \wedge i^- = j^- \wedge i^+ \leq j^+$. Hence any IA_c network can be translated to a PA_c network and so it is not surprising that both algebras have identical computational properties.

We can further exploit the above observation. Let us define a new algebra IA_p in the way that any constraint in IA_p can be defined as a conjunction of PA constraints. IA_p is called a *pointisible* subclass of interval algebra [116]. Clearly, any constraint from IA_c is also in IA_p but there are also other constraints in IA_p that are not convex. For example the constraint $\{b, o\}$ is not convex but it can be represented as $i^- < i^+ \wedge j^- < j^+ \wedge i^- < j^- \wedge i^+ < j^+ \wedge i^+ \neq j^-$ and hence it belongs to IA_p. In fact there exists only 187 constraints having the above property and belonging to IA_p. Again IA_p is closed under the composition and intersetion operations and because any

IA_p network can be translated to a PA network, the problem of deciding consistency in IA_p is tractable [116]. Obviously $IA_c \subset IA_p \subset IA$.

There exist other tractable subclasses of the interval algebra. The most studied subclass is a so-called *Ord-Horn Algebra* introduced in [85]. Let us first define the Ord-Horn constraints between the time points. An *Ord-Horn constraint* is a disjunction $c_1 \lor \ldots \lor c_k$ where at most one of the c_i constraints is an inequality constraint of the form $x \leq y$ and all other c_i constraints are of the form $x \neq y$, where x and y are variables over $\Re$. If i and j are intervals and R is a temporal constraint between them we call R a *Ord-Horn constraint* if $[i \ R \ j]$ can be equivalently expressed as a conjunction of Ord-Horn clauses on the end points of i and j. We can now define an interval algebra IA_H that contains only the Ord-Horn constraints. There are 868 such temporal relations satisfying the Ord-Horn property, which is more that 10% of all temporal constraints in IA. Moreover, this set is maximal in the sense that it cannot be extended without losing tractability [85]. There are exactly 18 maximal tractable subclasses of IA and reasoning on any other subclass of IA that is not included in these maximal subclasses is NP-complete [61].

p **before** i (i after p)	$p < i^-$	
p **starts** i (i started by p)	$p = i^-$	
p **during** i (i includes p)	$i^- < p, \ p < i^+$	
p **finishes** i (i finished by p)	$p = i^+$	
p **after** i (i before p)	$i^+ < p$	

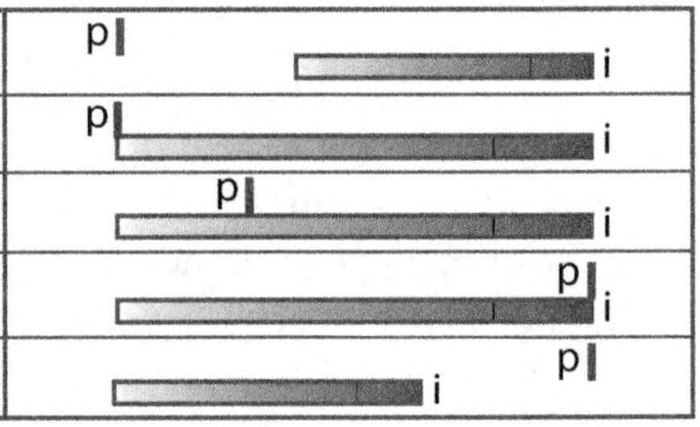

Figure 2.10: Primitive relations between a point and an interval.

In Section 2.1.2 we defined an interval by its two end points to establish an easy transition from the point algebra to the interval algebra. However, this is not a fully accurate approach to intervals. Assume the situation where we have a room with a light bulb that is off and after switching the toggle, the light becomes on. This situation can be described naturally using two intervals, *Off* for the time period when the light is off, and *On* for the time period when the light is on with the obvious relation *Off* $\{m\}$ *On*. Now, if we see the intervals as their end points then the end time of *Off* equals the start time of *On* (see Figure 2.4 for the definition of the *meet* relation). Let us denote this time point t. Then from the PA point of view the light bulb is both on and off at time t which is inconsistent. This shows that we need to see intervals as first-order objects and it is useful to define a formalism for reasoning both on points and intervals. In [73] the *qualitative algebra* (QA) for qualitative temporal reasoning was defined. In addition to point-to-point relations from PA (see Section 2.1.1) and interval-to-interval relations from IA the qualitative algebra also includes *point-to-interval* and *interval-to-point* relations. There are five of them as Figure 2.10 shows. The cited paper [73] goes further and it describes how to combine qualitative algebra with metric information. Jonsson and Krokhin [50] give a complete complexity classification of all subclasses of QA as either tractable or NP-complete.

2.1.4   HISTORICAL CONTEXT

Though we started this section with the description of the point algebra, it is the interval algebra that is in fact an older concept. It was the influential work of James Allen [2] that introduced the formalism for reasoning over time intervals—an *interval algebra*. Allen also proposed a path-consistency algorithm to handle it. Later on, Villain and Kautz [116] showed that this polynomial algorithm is not complete and they proved that deciding consistency of the interval algebra as well as computing its minimal network are NP-complete problems. In the same work they proposed the *point algebra* as a tractable alternative to the interval algebra and they identified the tractable *pointisible subclass* IA_p of the interval algebra which consists only of relations of IA that can be expressed as a conjunction of binary constraints in PA. Though the paper [116] claimed that path-consistency computes the minimal network for PA, Peter van Beek [113, 117] showed that this is true only for the subset PA_c of the point algebra containing the convex constraints (the constraint $\{<, >\}$ is omitted as it is not convex). Van Beek also proposed a strong 4-consistency algorithm to find minimal networks for PA and IA_p. This algorithm does not make the network globally consistent; as pointed out by Koubarakis [59] strong 5-consistency is necessary and sufficient for achieving global consistency for PA and IA_p. These works motivated further research in the area of tractable subclasses of IA. The *Ord-Horn Algebra* introduced in [85] is among the most studied. Recently Krokhin, Jeavons, and Jonsson [61] proved an important dichotomy result that classifies all maximal subclasses of IA as either tractable or NP-complete. In particular they showed that there are exactly 18 maximal tractable subclasses of IA and reasoning on any other subclass of IA that is not included in these maximal subclasses is NP-complete.

2.2   QUANTITATIVE TEMPORAL FRAMEWORKS

The second core approach to temporal reasoning uses metric temporal information of several types. We can express durations of events (*"I read newspapers for thirty minutes"*), temporal distance between the events (*"I entered the office one hour after finishing the breakfast"*), and absolute time of the events (*"I entered the office at 8:00 a.m."*). The temporal reference in quantitative reasoning is a time point (an interval is modeled using two time points) and the temporal constraints are quantitative, relative, and absolute numerical constraints.

Let us recall the situation from the section on Qualitative Temporal Frameworks and extend it with some metric temporal information.

> *"I got up at 6 o'clock. I read newspapers for thirty minutes during the breakfast. After the breakfast I walked to my office which took me one hour. I entered the office at 8:00 a.m."*

Again, we can ask whether the provided temporal information is consistent. This is a practically interesting task that appears in contexts, for example, when verifying evidence in crime investigation and when checking temporal validity of some story. We will present a formal consistency-checking procedure later, but for now we can verify the consistency of our story by checking that all the tasks—having breakfast, reading newspapers, and walking to office—fit between 6:00 a.m. and 8:00 a.m. As the following query-answer example will show, this is indeed possible.

In addition to checking consistency of temporal information, we can ask queries about the start or end of events, for example *"When did I start my breakfast?"*. Let us try to show how to answer this particular question. Similarly to qualitative reasoning, we will use the time points describing the starts and ends of all the activities in the situation, namely *bs* for the start of breakfast, *be* for the end of breakfast, *rs* and *re* for the start and end of reading newspapers, and finally *ws* and *we* for the start and end of the walking activity. Obviously, entering the office corresponds to the time point *we*—the end of walking. Figure 2.1 (b) shows the basic temporal relations between these variables, but we can now use more specific information from the description of the situation. Let us assume that the time-point variables describe the times of the events measured in minutes from midnight. Then we can specify the temporal constraints in the following way:

- $360 \leq bs$ expressing *"I got up at 6 o'clock"*, which actually means that *"I started breakfast not earlier than at 6:00 a.m.",*

- $bs < be$ stating that breakfast is an activity taking some non-zero time (this constraint is not necessary as the combination of the following three constraints gives a stronger temporal relation between bs and be, but in general, it should be included for completeness of the situation description),

- $bs \leq rs \wedge re \leq be$, i.e., *"I read newspapers during breakfast"*,

- $re - rs = 30$ expressing *"I read newspapers for thirty minutes"*,

- $be = ws$ expressing *"after breakfast I walked to my office"*,

- $we - ws = 60$ expressing *"[walking] took me one hour"*,

- $we = 480$ expressing *"I entered the office at 8:00 a.m."*

From these simple linear equality and inequality constraints we can easily deduce

$$bs \leq rs = re - 30 \leq be - 30 = ws - 30 = (we - 60) - 30 = 390.$$

Because we also know $360 \leq bs$ the answer to the query is $360 \leq bs \leq 390$, which is in words *"I started my breakfast between 6:00 a.m. and 6:30 a.m."*

This was an example of a so-called *Simple Temporal Problem* (STP). We will now formalize this problem, present some algorithms for checking consistency of an STP and for finding a minimal STP, and then we will extend this formalism to more general temporal constraints.

2.2.1   SIMPLE TEMPORAL PROBLEMS

Numerical temporal constraints describe absolute positions of time points and distances between pairs of time points. Assume that t_i and t_j are two time-point variables, that is, their possible values are real numbers. There are two types of *simple temporal constraints*:

- unary constraints of the form $a_i \leq t_i \leq b_i$ expressing the absolute position of the time point t_i,

- binary constraints of the form $a_{ij} \leq t_j - t_i \leq b_{ij}$ expressing the minimal (a_{ij}) and maximal (b_{ij}) distance between the time points t_i and t_j,

where a_i, b_i, a_{ij}, b_{ij} are real constants. Let t_0 be a special reference point located at time 0 ($t_0 = 0$). Then we can rewrite the unary constraint $a_i \leq t_i \leq b_i$ to the binary constraint $a_i \leq t_i - t_0 \leq b_i$. Hence, all constraints can be viewed as binary in the modeling framework. We will denote the simple temporal constraint between the time points t_i and t_j as a constraint $r_{ij} = [a_{ij}, b_{ij}]$. Notice that if $r_{ij} = [a_{ij}, b_{ij}]$ is a temporal constraint between the time points t_i and t_j then $r_{ji} = [-b_{ij}, -a_{ij}]$ is a symmetrical temporal constraint between the time points t_j and t_i. Let us notice that all the constraints from the above example situation are indeed simple temporal constraints. For example the constraint $we - ws = 60$ can be rewritten as $r_{ws,we} = [60, 60]$.

Similarly to qualitative approaches, we can define intersection and composition of simple temporal constraints. Note that the union of simple temporal constraints may give a constraint that is not simple temporal (it is defined using a union of intervals rather than a single interval). We will discuss the union of simple temporal constraints later in the text when defining a temporal constraint satisfaction problem.

- *Intersection* of constraints r_{ij} and r'_{ij} is a constraint $r_{ij} \cap r'_{ij} = [\max(a_{ij}, a'_{ij}), \min(b_{ij}, b'_{ij})]$ that corresponds to the conjunction of constraints $a_{ij} \leq t_j - t_i \leq b_{ij} \wedge a'_{ij} \leq t_j - t_i \leq b'_{ij}$.

- *Composition* of constraints r_{ij} and r_{jk} is a constraint $r_{ij} \circ r_{jk} = [a_{ij} + a_{jk}, b_{ij} + b_{jk}]$ that is obtained from the sum of constraints $a_{ij} \leq t_j - t_i \leq b_{ij}$ and $a_{jk} \leq t_k - t_j \leq b_{jk}$.

These two operations are important for the combination of constraints in the similar way as in qualitative reasoning. $r_{ij} \cap (r_{ik} \circ r_{kj})$ is a new constraint between the time points t_i and t_j that is obtained from the combination (intersection) of the existing constraint r_{ij} between the time points t_i and t_j and the induced constraint ($r_{ik} \circ r_{kj}$) describing the temporal distance via the time point t_k. This is the way we got the answer to the question *"When did I start my breakfast?"* in the above example. The ad-hoc reasoning process can now be expressed formally using the composition and intersection operations in the following way:

$$r_{bs,t_0} \cap ((r_{bs,be} \cap (r_{bs,rs} \circ r_{rs,re} \circ r_{re,be})) \circ r_{be,ws} \circ r_{ws,we} \circ r_{we,t_0})$$

which is

$$[-\infty, -360] \cap (([1, \infty] \cap ([0, \infty] \circ [30, 30] \circ [0, \infty])) \circ [0, 0] \circ [60, 60] \circ [-480, -480])$$

$$= [-390, -360].$$

There are several observations about the above example. Namely, we used some symmetrical constraints to compose them with other constraints, we used ∞ to express that some bound is not limited, and we expected breakfast to last for at least one minute.

Let us now formalize the Simple Temporal Problem and its properties.

Definition: [26] A *Simple Temporal Problem* (STP) consists of a set of n variables $X = \{t_1, t_2, \ldots, t_n\}$ representing the time points and a set C of simple temporal constraints between them. A *Simple Temporal Network* (STN) is a directed graph (X, C), where each arc from C is labeled by a constraint r_{ij}. (Usually only one of the equivalent constraints r_{ij} and r_{ji} is represented in the network.)

An STN is *consistent* if there exists an instantiation of variables X to real numbers satisfying all the constraints—a *solution* to the STP. An STN is *minimal* if every value in the interval r_{ij} (recall that a simple temporal constraint is represented as an interval) belongs to some solution, that is, for each $x_{ij} \in r_{ij}$ there is a solution to the STP such that $t_j - t_i = x_{ij}$.

Figure 2.11 (a) shows a simple temporal network describing our running example. We added the reference point t_0 and all the simple temporal constraints. This network is consistent but it is not minimal (for example the maximal distance between t_0 and bs is 390 minutes as we computed above).

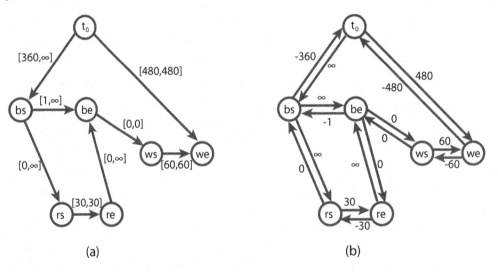

(a)   (b)

Figure 2.11: Example of a simple temporal network (a) and its distance graph (b).

A Simple Temporal Problem is a special instance of a Constraint Satisfaction Problem containing linear inequality constraints only. Such problems can be solved by the exponential simplex method [24] or by the polynomial, but rather complex Khachiyan's algorithm [54]. Hence the problem of deciding consistency of an STN is tractable. In fact, the simple temporal constraints are a special class of linear inequalities that can be solved in simpler ways as we shall show next.

An STN can be represented by a weighted directed acyclic graph, called a *distance graph* (the notion of *d-graph* is also used in literature), where the nodes correspond to time points and the directed arcs are annotated by weights in the following way. If there is a constraint $r_{ij} = [a_{ij}, b_{ij}]$ in the STN then there is a directed arc $t_i \rightarrow t_j$ annotated by weight b_{ij} and a directed arc $t_j \rightarrow t_i$ annotated by weight $-a_{ij}$ in the distance graph. In other words, each arc $t_i \rightarrow t_j$ annotated by weight w_{ij} in the distance graph represents a constraint $t_j - t_i \leq w_{ij}$, that is, a maximal distance between the time points. Figure 2.11 (b) gives an example of a distance graph. Now assume a path $t_i = t_{i_0}, t_{i_1}, \ldots, t_{i_k} = t_j$ in the distance graph. This path induces a constraint

$$t_j - t_i \leq \sum_{l=1}^{k} w_{i_{l-1}, i_l}.$$

As this constraint must hold for any path between the time points t_i and t_j, it is easy to verify that the constraint

$$t_j - t_i \leq d_{ij},$$

where d_{ij} is the length of the shortest path between t_i and t_j, must hold. Hence, to verify consistency of STNs it is useful to know the lengths of the shortest paths between all pairs of nodes. The Floyd-Warshall algorithm [21] is a popular all-pairs-shortest-path algorithm, which can be used to determine consistency of STPs. The algorithm can be described as follows:

```
for i, j = 1, ..., n do
    d_ii ← 0
    if (t_i, t_j) ∈ C then d_ij ← w_ij else d_ij ← ∞ fi
end
for k = 1, ..., n do
   for i, j = 1, ..., n do
      d_ij ← min(d_ij, d_ik + d_kj)
   end
end
```

Notice the similarity with the *path consistency* algorithm that we presented for the point algebra. The Floyd-Warshall algorithm can also be used for finding the minimal network for a given STN, and has time complexity $O(n^3)$ for the graph with n nodes.

Proposition: [105] An STP (X, C) is consistent if and only if its distance graph has no negative cycles.

If there is a negative cycle starting in node t_i then the sum of inequalities along the cycle yields $t_i - t_i < 0$, which cannot be satisfied. If there is no negative cycle in the distance graph then the shortest-path distances are well defined. In particular, the relation $d_{0j} \leq d_{0i} + w_{ij}$ holds for any pair of time points t_i and t_j and hence also $d_{0j} - d_{0i} \leq w_{ij}$. It means that the tuple $t_1 = d_{01}, \ldots, t_n = d_{0n}$ satisfies all the simple temporal constraints and is a solution to the STP ($t_0 = 0$ is the reference point in the corresponding STN). Consequently, the STN is consistent. There is another tuple, namely $t_1 = -d_{10}, \ldots, t_n = -d_{n0}$, that naturally satisfies all the temporal constraints and is also a solution to a given STP. This is because $d_{i0} \leq w_{ij} + d_{j0}$, which yields $-d_{j0} - (-d_{i0}) \leq w_{ij}$. Interestingly, the numbers $-d_{i0}$ and d_{0i} define the earliest and latest possible times for the time point t_i. In fact, if we design an STN, where the constraints between the time points t_i and t_j are defined as $[-d_{ji}, d_{ij}]$, then we obtain a minimal STN.

2.2.2   TEMPORAL CONSTRAINT SATISFACTION PROBLEMS

Assume now a slightly extended situation from the previous section, where one more person is involved:

> "I got up at 6 o'clock. I read newspapers for thirty minutes during breakfast. After breakfast I walked to my office which took me one hour. I entered office exactly at the same time as Peter who left his home at 7 a.m. Peter is going to office either by a car, which takes him 15-20 minutes, or by a bus, which takes 40-50 minutes."

In this situation the time of entering office is not given and it is not known which means of transport Peter used. Moreover, such a situation cannot be modeled using a simple temporal network. In particular, let ps be the time when Peter started his trip to the office and pe be the time when he arrived to the office. From the situation description we know the following two constraints:

- $we = pe$ expressing that I arrived to the office at the same time as Peter,

- $15 \leq pe - ps \leq 20 \vee 40 \leq pe - ps \leq 50$ describing the two possible options for Peter's transport.

It is the disjunctive constraint that cannot be expressed directly in a simple temporal network and we need a more general framework known as a *Temporal Constraint Satisfaction Problem* (TCSP) [26]. A binary constraint in a TCSP is expressed using a set of intervals rather than a single interval as in an STP. In particular, a binary constraint $T_{ij} = \{[a_1, b_1], [a_2, b_2], \ldots [a_k, b_k]\}$ between time points t_i and t_j represents the disjunction:

$$a_1 \leq t_j - t_i \leq b_1 \vee a_2 \leq t_j - t_i \leq b_2 \vee \ldots \vee a_k \leq t_j - t_i \leq b_k.$$

This set of intervals can be seen as a union of simple temporal constraints over the same pair of temporal variables. We may assume that the intervals are pairwise disjoint in a canonical form. Obviously, if two intervals intersect then we can substitute them by their union and by repeating this process until obtaining disjoint intervals we will get the canonical form. For example, if $T_{ij} = \{[1, 5], [3, 7], [9, 12]\}$ then the canonical form is $T_{ij} = \{[1, 7], [9, 12]\}$.

We can define the three operations on temporal constraints: union, intersection, and composition, that can be used in inference and constraint propagation algorithms. Let $T = \{I_1, \ldots, I_m\}$ and $R = \{J_1, \ldots, J_n\}$ be two temporal constraints, where $I_1, \ldots, I_m, J_1, \ldots, J_n$ are intervals (simple temporal constraints).

- If T and R are defined over the same pair of temporal variables, then we define the *union* of T and R, denoted as $T \cup R$, as $\{I_1, \ldots, I_m, J_1, \ldots, J_n\}$. Note that the obtained constraint is not necessarily in the canonical form, but as we showed above, it can be easily transformed to the canonical form. The union of temporal constraints corresponds to their disjunction.

- If T and R are defined over the same pair of temporal variables, then we define the *intersection* of T and R, denoted as $T \cap R$, as a set $\{K_1, \ldots, K_r\}$, where each K_i is obtained by intersection of some intervals I_k and J_l. Some of these intersections might be empty and hence $r \leq m + n$. If the original constraints were in canonical forms and the empty sets are removed from $T \cap R$ then we obtain a constraint in a canonical form. The intersection of temporal constraints corresponds to their conjunction.

- Finally, if T is a temporal constraint between time points t_a and t_b and R is a temporal constraint between time points t_b and t_c then the *composition* of constraints T and R, denoted as $T \circ R$, gives an induced constraint between t_a and t_c defined as a set $\{K_1, \ldots, K_r\}$, where each K_i is obtained as a composition of simple temporal constraints I_k and J_l. Recall that if $I_k = [x, y]$ and $J_l = [u, v]$ then $I_k \circ J_l = [x + u, y + v]$. As some obtained intervals may overlap, we obtain at most $m.n$ intervals in the composed constraint.

Going back to our example situation, we can graphically describe it using a temporal network where nodes correspond to the time points and arcs are annotated by sets of intervals describing particular constraints as Figure 2.12 shows.

If we now ask the question *"How does Peter arrive to the office?"* we can find the answer using the following reasoning process. Notice first that this question is not asking about the temporal information directly but if we deduce that one of the alternatives *"going by car"* and *"going by bus"* is not consistent with the rest of the temporal network then we can indeed answer the question. This is exactly what we shall show now.

We can infer a new constraint $T'_{t_0,pe}$ between t_0 and pe by composing the known constraints $T_{t_0,bs}, T_{bs,rs}, T_{rs,re}, T_{re,be}, T_{be,ws}, T_{ws,we}, T_{we,pe}$. These are simple temporal constraints so the composition is simple. We get $T'_{t_0,pe} = [450, \infty]$. Similarly, we can compose the constraints $T_{t_0,ps}$ and $T_{ps,pe}$ to obtain $T''_{t_0,pe} = \{[435, 440], [460, 470]\}$. As the two new constraints are between the same time points we can make their intersection to get a constraint $T_{t_0,pe} = T'_{t_0,pe} \cap T''_{t_0,pe} = [460, 470]$. By the intersection of the constraint $T_{ps,pe}$ with the composition of constraints $T_{ps,t_0} = [-420, -420]$ and $T_{t_0,pe} = [460, 470]$ we get a new tighten constraint $T_{ps,pe} = [40, 50]$ between the time points ps and pe, which corresponds to *"going by bus"* (the option *"going by car"* is excluded). Hence the answer to the above question is *"Peter used a bus to go to the office."*

We define formally a Temporal Constraint Satisfaction Problem as follows:

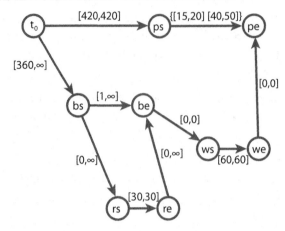

Figure 2.12: Example of a temporal network.

Definition: [26] A *Temporal Constraint Satisfaction Problem* (TCSP) consists of a set of variables $X = \{t_1, \ldots, t_n\}$ with continuous domains representing time points and a set of binary temporal constraints, where each constraint is represented by a set of intervals $\{[a_1, b_1], [a_2, b_2], \ldots [a_k, b_k]\}$.

A TCSP is *consistent* if there exists an instantiation of variables X to real numbers satisfying all the constraints—a *solution* to the TCSP. We say that a value v is *feasible* for variable t_i, if there exists a solution to the TCSP where $t_i = v$. The constraint T_{ij} between t_i and t_j is *minimal* if its representation consists of all feasible values for $t_j - t_i$.

A *Temporal Network* is a directed graph (X, C), where each arc from C is labeled by a respective constraint T_{ij}. The temporal network is *minimal* if all the constraints are minimal.

A straightforward way to solve a TCSP is to decompose it into several simple temporal problems, that can be solved in a polynomial time, and then composing their solutions. Recall that a temporal constraint is represented by a set of intervals—a disjunction of simple temporal constraints between the same time points. Selecting one disjunct for each temporal constraint gives an STP called a *component* STP. Hence a TCSP can be seen as a collection of component STPs. Figure 2.13 shows the component STPs for the temporal network from Figure 2.12.

It is easy to see that a TCSP is consistent if and only if at least one of its component STPs is consistent. Moreover, the minimal temporal network can be computed from the minimal networks of its component STPs as stated in the following theorem.

Theorem: [26] The minimal network N for a given TCSP is a union of minimal networks of its component STPs. Formally, $N = \cup_i N_i$, where N_i are minimal networks for all component STPs.

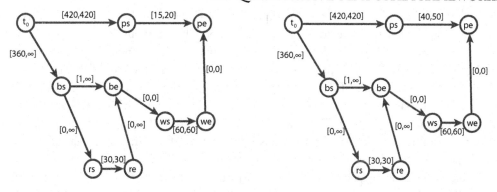

Figure 2.13: Component STPs for the temporal network from Figure 2.12.

Figure 2.14 shows the minimal network for the Temporal Constraint Satisfaction Problem from Figure 2.12.

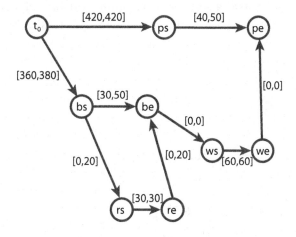

Figure 2.14: Minimal temporal network for the network from Figure 2.12.

As the complexity of finding a minimal network for an STP with n nodes is $O(n^3)$, the complexity of finding the minimal network by generating and solving all component STPs is $O(n^3 k^e)$, where e is the number of edges and k is the maximum number of intervals in the temporal constraints. It seems that the exponential time complexity is inevitable as the problem of verifying consistency of a TCSP is NP-complete [26]. Nevertheless, we can find algorithms with significantly better runtime than the runtime of the above brute-force algorithm. The idea is representing the problem as a *meta-CSP* (roughly, a CSP whose domains are CSPs) and solving the meta-CSP by a combination of backtracking search with pruning techniques [102]. A temporal

CSP is formulated as follows. For each temporal constraint (arc in the temporal network) we define a decision variable with the domain consisting of possible intervals in the temporal constraint. The backtracking algorithm assigns intervals to arcs which corresponds to adding a new arc to a simple temporal network. The algorithm then verifies that the obtained STP does not contain a negative cycle (the current STP is consistent). If this condition holds, the algorithm continues with the next variable (arc). Otherwise, the algorithm backtracks, which means that it tries to assign a different interval to the last variable (arc). Notice that the meta-CSP does not contain explicit constraints, but the consistency of assignments is defined implicitly via the STP defined by the selected intervals—simple temporal constraints.

Recall that to solve qualitative temporal reasoning problems we used path-consistency techniques. We can also apply path consistency to a TCSP as well, for example to preprocess the temporal network by removing some inconsistent values.

Definition: We say that a constraint T_{ij} is *path consistent* if and only if $T_{ij} \subseteq \cap_k (T_{ik} \circ T_{kj})$. The temporal network is path consistent iff all its constraints are path consistent.

The general path consistency algorithms can be naturally reformulated for temporal constraints. We will present here the algorithm based on the widely used PC-2 algorithm by Mackworth [68]. The version of PC-2 algorithm for temporal networks is also called NPC-2.

$Q \leftarrow \{(i, k, j) : i < j, k \neq i, j\}$
while $Q \neq \emptyset$ do
$\quad Q \leftarrow Q - \{(i, k, j)\}$
$\quad T \leftarrow T_{ij}$
$\quad T_{ij} \leftarrow T_{ij} \cap (T_{ik} \circ T_{kj})$
$\quad$ if $T_{ij} = \emptyset$ then $return(INCONSISTENT)$ fi
$\quad$ if $T \neq T_{ij}$ then $Q \leftarrow Q \cup \{(i, j, k) : i < k, k \neq j\}$
$\quad\quad\quad\quad\quad\quad\quad \cup \{(k, i, j) : k < j, k \neq i\}$
$\quad\quad\quad\quad\quad\quad\quad \cup \{(j, i, k) : j < k, k \neq i\}$
$\quad\quad\quad\quad\quad\quad\quad \cup \{(k, j, i) : k < i, k \neq j\}$ fi
end

Since the temporal variables are defined over the real numbers, the termination of the NPC-2 algorithm is not obvious. However, if the TCSP is integral (extreme points of the intervals are integers) then it is easy to show that the algorithm terminates as the constraints are tightened by an integral amount. For a constraint $T_{ij} = \{[a_1, b_1], [a_2, b_2], \dots [a_k, b_k]\}$ we define its *range* as $b_k - a_1$. Let R be the maximum range over all constraints, called a *range* of the network. Then the algorithm NPC-2 achieves path consistency in $O(n^3 R)$ composition steps. However, as pointed in [102] enforcing path consistency on TCSPs is problematic when the range is large. Recall that the upper bound on the number of intervals in the composition operation is $m.n$, where m and n are the numbers of intervals in the composed constraints. As the composition operation is repeated many times, the total number of intervals in the path-consistent network might be exponential relative to the number of intervals per constraint in the input network. This is called a *fragmentation problem*. Figure 2.15 gives a particular example showing how the number of intervals increases when the network is made path consistent. The fragmentation problem can be avoided by enforcing looser consistency than path consistency. Schwalb and Dechter [102] proposed one such algorithm called Upper-Lower-Tightening (ULT).

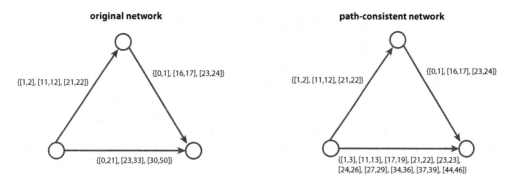

Figure 2.15: Example of the fragmentation problem [102].

2.2.3 DISJUNCTIVE TEMPORAL PROBLEMS

In the example situation in the previous chapter, we knew that Peter left home at 7 a.m. Assume now that we vary the information in the problem specification as follows:

> *"Peter got up at 6 o'clock and before leaving home he went jogging for 40 minutes and had a breakfast which took him 20 minutes. Peter is going to office either by a car, which takes him 15–20 minutes, or by a bus, which takes 40–50 minutes."*

This information can be encoded in a temporal network as depicted in Figure 2.16. We use there the following time points:

- *pbs* and *pbe* to denote the start time and end time of Peter's breakfast,

- *pjs* and *pje* to denote the start time and end time of Peter's jogging, and

- *ps* to denote the time when Peter left home and started his trip to the office and *pe* denoting the time when Peter entered the office.

Obviously, the information in this temporal network is not fully accurate. If we construct the minimal network we may notice that the minimal time when Peter can leave home is 6:40 a.m. (400 minutes from the midnight). This is because the provided information allows two activities—jogging and having breakfast—to overlap in time, which is usually not the case in human life.

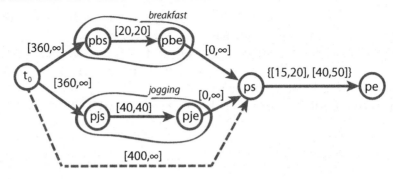

Figure 2.16: Temporal network with overlapping activities "breakfast" and "jogging".

The situation description does not give additional information about whether Peter had breakfast before jogging or vice versa and hence we cannot add a particular ordering constraint between the time points describing these activities. Nevertheless, we can enforce the constraint that these two activities do not overlap in time by saying that either breakfast finished before jogging started or jogging finished before the start of breakfast. The following disjunctive constraint expresses this relation:

$$pbe \leq pjs \vee pje \leq pbs.$$

This is a typical form of describing a so-called *unary resource*, a resource that can service at most one activity at any time. Though this is a disjunctive constraint over temporal variables, it is not the temporal constraint according to the specification of a TCSP. The temporal constraints in a TCSP are always binary; they express possible distances between two time points. The above constraint is a 4-ary constraint and hence a more generalized formalism is necessary to describe such situations. This formalism is known as a *disjunctive temporal problem* (DTP). Briefly speaking, a TCSP allows disjunctions of simple temporal constraints that are defined over the same pair of temporal variables while a DTP allows disjunctions of arbitrary simple temporal constraints even if these constraints are defined over different pairs of temporal variables.

Definition: [107] A *Disjunctive Temporal Problem* (DTP) consists of a set of variables $X = \{t_1, \ldots, t_n\}$ having continuous domains and representing time points and a set of disjunctive difference constraints between the time points in the form:

$$a_1 \leq x_1 - y_1 \leq b_1 \vee a_2 \leq x_2 - y_2 \leq b_2 \vee \ldots \vee a_k \leq x_k - y_k \leq b_k,$$

where $x_1, y_1, \ldots x_k, y_k$ are time points from X and $a_1, b_1, \ldots, a_k, b_k$ are real numbers.

A DTP is *consistent* if there exists an instantiation of variables X to real numbers satisfying all the constraints—a *solution* to a DTP.

As Disjunctive Temporal Problems are very close to Temporal Constraint Satisfaction Problems, the same core solving approach can be used to verify consistency of a DTP. Recall that this approach is based on two steps:

- decompose the DTP to a set of component simple temporal problems; each STP contains exactly one simple temporal constraint from each disjunctive difference constraint in the DTP,

- check consistency of these STPs.

Stergiou and Koubarakis [107] were the first to study various backtracking algorithms to solve the meta-CSP with variables corresponding to disjunctions. They presented the theoretical results characterizing the algorithms in terms of search nodes visited and consistency checks performed. Armando et al. [4] used SAT techniques for the selection of constraints from the disjunctions. As a DTP is a generalization of a TCSP, the problem of deciding consistency of a DTN is NP-complete.

2.2.4   TEMPORAL NETWORKS WITH ALTERNATIVES

Examples in previous chapters introduced alternatives to temporal networks. These alternatives were expressed as disjunctions of temporal constraints. However, all the time points in the temporal networks were always assumed to be present and the alternatives only described optional temporal relations between them. In some problems it may be useful to express explicitly the existence of a time point. In our example situation we had two activities *"going by car"* and *"going by bus"* which cannot both occur. We modeled the situation as a single activity *"going to work"* with a variable duration depending on means of transport, with time points *ps* and *pe* modeling the start and end times of the activity. However, in some cases we may need a direct access to the choice of activity, for example to express the situation

> *"When Peter goes by a car then Robert joins him otherwise Robert goes by train which takes him 45 minutes."*

In such problems we need to express separately the time points when Peter and Robert left their homes and arrived to the office, even though there is a connection between them, but only when they both use Peter's car. A natural solution is at hand by introducing "shared" time points and enforcing a constraint that only some time points appear in the solution network. To model such problems Barták and Čepek [8] introduced a concept of *temporal network with alternatives* (TNA). The general idea is that each time point is annotated by a so-called *validity* variable that indicates whether the point is present in the solution or not. In addition to simple temporal constraints between the time points (in which case the framework is called a simple temporal network with alternatives), there are logical constraints between the validity variables restricting which time points can be present together in the solution. These logical constraints are connected to so-called *fan-in* and *fan-out* subgraphs of the network and describe a type of branching; hence they are called *branching constraints*. Figure 2.17 shows a temporal network with alternatives for the above situation. For each person there is an alternative branching (going either by train or by car for Robert and going either by car or by bus for Peter) and there is a parallel branching for going by car (then both Peter and Robert use a car).

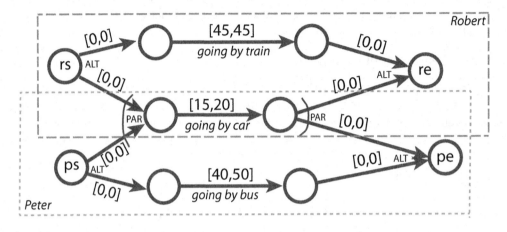

Figure 2.17: A Temporal Network with Alternatives modeling alternative activities.

The motivation behind TNAs arose from manufacturing processes that can split to parallel and alternative sub-processes. More formally, assume that node x has direct successors $y_1, \ldots, y_k$ in the graph, that is, there are arcs (x, y_i) in the graph. Then *parallel output branching* in node x means that either all nodes $x, y_1, \ldots, y_k$ are in the solution (in the process) or none of these nodes is there. *Alternative output branching* in node x means that either node x and exactly one of the nodes $y_1, \ldots, y_k$ are in the solution or none of these nodes is there. The input branching is defined in a similar way. A typical task is to select the nodes while respecting the branching and temporal constraints. For example, if we know that it took less than 30 minutes for Peter to reach his office (this is an extra temporal constraint connecting points *ps* and *pe*) then we can deduce that both Peter and Robert went by car and the alternatives "going by train" and "going by bus" are discarded (the related temporal constraints are not assumed).

Definition: [8] A *Simple Temporal Network with Alternatives* (STNA) is a directed acyclic graph, where nodes correspond to time points and arcs describe simple temporal relations between the time points. Each node is annotated by the type of input and output branching (parallel and alternative branching can be used). An STNA is *consistent* if there exists a sub-graph that satisfies the branching constraints and this sub-graph forms a consistent STN.

By conversion from the 3-SAT problem, it is possible to show that the problem of deciding consistency of an STNA is NP-complete [8]. This result holds even if precedence constraints are used instead of simple temporal constraints (a precedence constraint is a constraint in the form

$c \leq t_i - t_j$, where c is a constant). Therefore a specific form of a TNA was proposed—a so-called *Nested TNA*—where the problem of deciding consistency is tractable for precedence constraints and becomes NP-complete when simple temporal constraints are used [9]. Nested TNAs are graphs with a specific structure obtained by applying alternative and parallel decomposition operations. Briefly speaking, for each node, where the process splits, there is a downstream node, where the process joins. The TNA from Figure 2.17 is not a Nested TNA (the possible nests interleave there) while the TNA in Figure 2.18 is nested. Nested TNAs resemble the structure of hierarchical task networks [99] and the nested structure appears in many real-life processes [7].

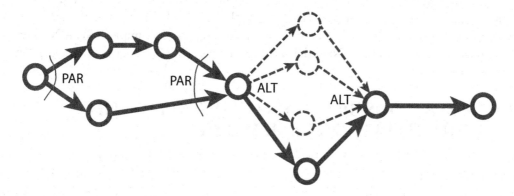

Figure 2.18: Example of a Nested Temporal Network with Alternatives (one selected process is highlighted using the bold lines while the dashed lines show the discarded alternatives).

The algorithms for verifying consistency of (Nested) TNAs use a similar two-step process as solving TCSPs and DTPs. In one step, the nodes are being selected and in the second step, the consistency of the obtained STN is being verified [9].

2.2.5   HISTORICAL CONTEXT

The first work on metric temporal reasoning relied on linear constraints solved using linear programming techniques [70]. One of the most influential works that introduced the concept of quantitative temporal reasoning and put it within the context of constraint processing was the seminal paper [87]. This paper defined a Temporal Constraint Satisfaction Problem (TCSP) and its tractable sub-class—a Simple Temporal Problem (STP). Path consistency algorithms were studied there and the Floyd-Warshall's all-pairs-shortest-path algorithm was used to solve an STP in polynomial time. Since then many algorithms improving practical efficiency of checking consistency of an STP were proposed [90].

An STP is still the most widely used framework for quantitative temporal reasoning, thought more expressive frameworks were proposed such as a Disjunctive Temporal Problem (DTP) [107]. A DTN extends a TCSP by allowing disjunctions of arbitrary simple temporal constraints rather than using simple temporal constraints between the same variables as in a TCSP. DTPs and TCSPs are typically solved using meta-CSP formulations [4, 102]. In addition to consistency enforcing algorithms there is a related research of tractability of metric temporal CSPs. It is known that a TCSP and hence also a DTP are NP-complete problems [26]. Koubarakis [59] showed that if simple temporal constraints are used together with disequations in the form $x - y \neq r$, where r is a real constant, then the problem of deciding consistency of such a temporal network is still tractable. The paper [59] has also shown that strong 5-consistency is the necessary and sufficient condition for achieving global consistency for such networks.

Another extension of an STP that adds specific logical branching constraints is called a Simple Temporal Network with Alternatives (Simple TNA) [8]. This branch of research is motivated by modeling workflows [7]. Again, deciding consistency of a Simple TNA is NP-complete [8], a tractable subclass called a Nested Temporal Network with Alternatives (Nested TNA) was proposed [9].

2.3   RELATIONS BETWEEN QUALITATIVE AND QUANTITATIVE FRAMEWORKS

In this section we will explore the relations between qualitative and quantitative frameworks and we will talk about the means to combine them. It is easy to see that a constraint network in the point algebra is a special case of a TCSP with specific metric constraints. We need to assume open and semi-open intervals, namely $(0, \infty)$ (an interval of positive numbers without zero) and $[0, \infty)$ (an interval of non-negative numbers, that is, positive numbers and zero). This extension does not change the TCSP framework, but we need to modify the union and intersection operations accordingly. We can then express the *PA* constraints in the following way:

- $t_i < t_j$ as a temporal constraint $T_{ij} = \{(0, \infty)\}$

- $t_i \leq t_j$ as a temporal constraint $T_{ij} = \{[0, \infty)\}$

- $t_i = t_j$ as a temporal constraint $T_{ij} = \{[0, 0]\}$

- $t_i \neq t_j$ ($t_i\{<, >\}t_j$) as a temporal constraint $T_{ij} = \{(-\infty, 0), (0, \infty)\}$

- $t_i > t_j$ as a temporal constraint $T_{ij} = \{(-\infty, 0)\}$

- $t_i \geq t_j$ as a temporal constraint $T_{ij} = \{(-\infty, 0]\}$

Consequently, reasoning in *PA* can be fully realized by reasoning in a *TCSP*. Moreover, notice that only the relation $\neq$ needs to be represented by a disjunctive constraint; all other *PA* relations are represented by simple temporal constraints. The inequality constraint indeed plays an exceptional role. For example, recall that the path-consistency algorithm computes the minimal network for PA_c (*PA* algebra without the relation $\neq$), whereas this is no longer true for the full *PA* as Figure 2.3 shows.

Though we can translate the primitive IA relations to simple temporal constraints, some combined IA relations such as $\{b, bi\}$ require non-binary temporal constraints ($x^+ < y^- \lor y^+ < x^-$ for x $\{b, bi\}$ y). Hence IA networks cannot always be translated into binary TCSPs. Nevertheless, this translation can be mapped to a DTP that supports n-ary disjunctive constraints.

There has been research involving the combination of qualitative and quantitative approaches. Meiri [73] combined the qualitative algebra with temporal constraints from a TCSP to obtain *general temporal constraint networks* where nodes are points and intervals and constraints are taken from the qualitative algebra and from a TCSP. To illustrate, Meiri considered the following scenario:

> "John and Fred work for a company that has local and main offices in Los Angeles. They usually work at the local office, in which case it takes John less than 20 minutes and Fred 15–20 minutes to get to work. Twice a week John works at the main office, in which case his commute to work takes at least 60 minutes. Today John left home between 7:05–7:10 a.m., and Fred arrived at work between 7:50–7:55 a.m. We also know that Fred and John met at a traffic light on their way to work."

Figure 2.19 shows a general temporal constraint network representing this scenario. The time t_0 now corresponds to 7:00 a.m., temporal variables js and je represent the start and end of John's travel to work, which itself is represented as an interval $J = (js, je)$, and similarly fs and fe represent the start and end of Fred's travel to work ($F = (fs, fe)$). Queries that can be answered by this network include *"Is the information in this story consistent?"*, *"Who was the first to arrive at work?"*, and *"What are the possible times at which John arrived at work?"*. Meiri [73] proposed a brute-force search algorithm to solve such problems and also studied path-consistency algorithms for computing the approximation of the minimal network.

Kautz and Ladkin [53] proposed a similar framework combining interval algebra with simple temporal constraints over the endpoints of intervals. They also presented translation algorithms between the metric and non-metric sub-languages and proposed a constraint-propagation procedure for the combined framework. Yet another combined framework has been proposed by Krokhin et al. [62]. This framework uses intervals and IA qualitative constraints combined with DTP constraints over the interval endpoints so it is a generalization of the framework by Kautz and Ladkin. Finally, Krokhin et al. [62] precisely characterize all subproblems that are tractable; all remaining problems are shown to be NP-complete.

2.4 SUMMARY

We presented two core frameworks of qualitative (non-metric) temporal reasoning, namely a *point algebra* (PA) [116] and an *interval algebra* (IA) [2], and also their combination called a *qualitative algebra* (QA) [73]. The point algebra uses time points as temporal references, the interval algebra uses intervals as temporal references, and the qualitative algebra can use both time points and intervals. Specific sub-classes such as PA_c, IA_c, IA_p exist. PA_c is a point algebra with convex con-

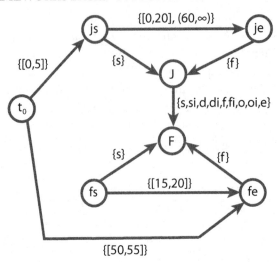

Figure 2.19: Example of a General Temporal Constraint Network.

straints only, i.e., without the constraint $\{<, >\}$. Similarly IA_c is an interval algebra with convex constraints only. IA_p is a pointisible subclass of IA, where the constraints are conjunctions of the PA constraints. The problem of deciding consistency of constraint networks is tractable for PA, PA_c (path consistency gives a minimal network there), IA_c, and IA_p while the same problem is NP-complete for IA.

Quantitative (metric) temporal frameworks use time points only and constraints represent restricted metric distances between the time points there. We described *simple temporal problems* (STP) and *temporal constraint satisfaction problems* (TCSP) [26] modeling binary temporal relations and *disjunctive temporal problems* (DTP) [107] supporting n-ary disjunctive constraints. An STP supports simple temporal constraints in the form $a_{ij} \le t_j - t_i \le b_{ij}$, a TCSP supports disjunctions of simple temporal constraints over the same pair of temporal variables (binary constraints), and finally a DTP supports disjunctions of arbitrary simple temporal constraints (n-ary constraints). There also exist frameworks combining temporal constraints with logical constraints (*temporal networks with alternatives* (TNA) [8]), with resource constraints (*resource constraint networks* [63]), and with finite domain constraints [74].

Figure 2.20 gives an overview of presented temporal frameworks and their main properties. For each framework we include its acronym, name, type of reasoning (qualitative, quantitative), used temporal reference and allowed constraints, and finally the complexity class of the problem of deciding about consistency of constraint networks.

	name	approach	temporal reference	temporal propositions	complexity
PA	point algebra	qualitative	time points	$\{<, =, >\}$	tractable
IA	interval algebra	qualitative	intervals	$\{b, m, o, s, d, f, e, bi, mi, oi, si, di, fi\}$	NP-c
QA	qualitative algebra	qualitative	intervals, time points	interval-to-point, IA, PA	NP-c
STP	simple temporal problem	quantitative	time points	binary difference	tractable
TCSP	temporal constraint satisfaction problem	quantitative	time points	disjunctive binary difference	NP-c
DTP	disjunctive temporal problem	quantitative	time points	n-ary disjunctive difference	NP-c
TNA	temporal network with alternatives	quantitative	time points	precedence, logical (branching)	NP-c
	general temporal constraint satisfaction problem	qualitative, quantitative	intervals, time points	TCSP, QA	NP-c

Figure 2.20: An overview of main temporal frameworks.

CHAPTER 3

Extensions: Preferences and Uncertainty

We will now consider how the formalisms described in the previous chapter can be extended. Although remarkably expressive, none of the approaches we have seen so far allow to distinguish among solutions according to preferences we may have on how constraints should be satisfied. Moreover, all the variables are assumed to be under the control of the executing agent and no uncertainty on when events occur or on which events occur is taken into account. The extensions we will now describe tackle precisely these limitations.

As we will see, the fundamental components underlying a time-aware rational agent are the same as in the underlying quantitative and qualitative approaches. The *temporal knowledge base* is extended with preferential information and with details concerning uncertainty. Procedures for checking *consistency* are replaced by procedures which enforce optimality and/or robustness to uncertainty. New *inference mechanisms* propagate preferences as well as changes tied to the presence of uncontrollable events.

3.1 PREFERENCES

While many frameworks for reasoning about time rely on the assumption that, along as precedences are respected and there are no overlapping activities, any available time is just as good, all of us know that that is really not true! All of us would prefer avoiding heavy meetings right after lunch, waiting for connecting flights for endless hours, or going on vacation at a selected destination during its rainy season. Among the many solutions to a temporal problem, being able select most preferred ones is a capability which is desirable in any intelligent system. On the other side, we may be often faced with over-constrained problems where relaxing hard constraints in a smart way can allow us to find a solution representing a better compromise.

We will now show how an efficient and expressive way to reason about preferences can help to handle time in a more flexible and sophisticated way. Among the many frameworks for temporal reasoning, constraint-based ones (quantitative and qualitative) have provided the most suitable base for the introduction of preferences. We will discuss how temporal constraints have been extended to allow for the representation of preferences, and, later, how temporal preferences can coexist with uncertainty and how they can be made conditional. Moreover, we will show how preferences allow for a significant increase in terms of representational power often at a modest additional computational cost.

3.1.1   PREFERENCES IN QUALITATIVE FRAMEWORKS

Fuzzy qualitative temporal reasoning [6, 32, 84, 86] can safely be considered the first extension to preferences of a constraint-based temporal model.

The main result in this line of research is the fuzzyfication of Allen's interval algebra (see Section 2.1) [6], denoted with IA^{fuz}. Specifically, constraints are extended by allowing not only to specify which subset of Allen's primitive relations should be considered but also by associating to each relation a degree of preference. Thus a constraint in this model has the form

$$i_1(r_1[\alpha_1], r_2[\alpha_2], \ldots)i_2$$

where i_1 and i_2 are two temporal intervals, r_i, for $i = 1, 2, \ldots$, is a primitive relation and α_i, for $i = 1, 2, \ldots$, is a preference degree in $[0, 1]$. Given a constraint R and a primitive relation r_i the preference degree assigned by R to r_i is denoted with $deg_R(r_i)$. The semantics induced by this syntax is that a primitive relation with degree α is a subset of $\mathbb{R}^2 \times \mathbb{R}^2$ to which the pairs of intervals which satisfy the primitive relation in a "classical" sense belong with membership degree α and all other pairs with membership degree 0. The semantics of a general constraint (that is, of a subset of the 13 primitive relations) is, then, the union of the fuzzy subsets corresponding to each of the relations. The fuzzy algebra IA^{fuz} is thus defined on the set

$$I = \{b[\alpha_1], m[\alpha_2], o[\alpha_3], s[\alpha_4], d[\alpha_5], f[\alpha_6], e[\alpha_7], bi[\alpha_8], mi[\alpha_9], oi[\alpha_{10}], si[\alpha_{11}], di[\alpha_{12}], fi[\alpha_{13}]\}$$

where $\alpha_i \in [0, 1]$, for $i = 1, \ldots, 13$, and a IA^{fuz} network is a set of constraints as defined above.

The operations defined on IA (see Section 2.1) can naturally be extended to constraints in IA^{fuz}. In particular, *inversion*, as expected, only affects relations and not preferences. Moreover, given two fuzzy relations $r = (r_1[\alpha_1], r_2[\alpha_2], \ldots, r_{13}[\alpha_{13}])$ and $r' = (r_1[\alpha'_1], r_2[\alpha'_2], \ldots, r_{13}[\alpha'_{13}])$, their *conjunction* (generalizing intersection), is defined as

$$r \otimes r' = (r_1[\beta_1], r_2[\beta_2], \ldots, r_{13}[\beta_{13}])$$

where $\beta_i = \min(\alpha_i, \alpha'_i)$, their *disjunction* (generalizing union) is defined as

$$r \oplus r' = (r_1[\beta_1], r_2[\beta_2], \ldots, r_{13}[\beta_{13}])$$

where $\beta_i = \max(\alpha_i, \alpha'_i)$ and their composition is defined as

$$r \circ r' = (r_1[\beta_1], r_2[\beta_2], \ldots, r_{13}[\beta_{13}])$$

where $\beta_i = \max_{j,k:r_i \in \{r_j \circ r_k\}} \min(\alpha_j, \alpha'_k)$, $i, j, k \in \{1, 2, \ldots, 13\}$.

The introduction of degrees in the constraints induces a grading of local consistency as well. In particular, if, as in the Interval Algebra IA case described in Section 2.1.2, we call a *singleton labeling* the choice of a primitive constraint on each constraint, its *degree of consistency* is defined as the minimum degree associated to any of its primitive relations. At this point, for clarity we should note that, as for the classical case (see Section 2.1.2) a *solution* of a IA^{fuz} network is an assignment of an interval (defined by a pair of real numbers representing, respectively, the lower and upper bound) to each node and that each solution identifies a singleton labeling. However, the consistency degree depends only on the singleton labeling and, thus, solutions identifying the same labeling have the same consistency degree.

Given an IA^{fuz} network N it is said to be *minimal* if and only if, for every constraint $r_{ij} \in N$ defined on intervals i_i and i_j, and $\forall r_k[\alpha] \in r_{ij}$ there is a complete singleton labeling s which assigns r_k to r_{ij} and such that its consistency degree is α.

Once an IA^{fuz} network is defined, there are several interesting questions one may want to answer: for example, determining the consistency degree of the network (i.e., the degree assigned to its optimal solutions), or finding an optimal singleton labeling, or computing the minimal network.

Such interesting questions have been answered in the literature in two main ways: by extending IA algorithms or by more general approaches, such as Branch and Bound.

As seen in Section 2.1, in the context of IA, minimality is enforced on a network by means of constraint propagation. In [6] the classical path consistency algorithm (see Section 2.1) is generalized to networks in IA^{fuz} yielding a complexity of $O(kn^3)$ where n is the number of intervals and k is the total number of different degrees which appear somewhere in the network.

As for IA also its sub-algebras have been extended to allow for preferences. In particular, the following have been defined and proven to be algebras [5]:

- Fuzzy extensions of classical point algebras PA and PA_c (see Sections 2.1.1 and 2.1.3):

  - PA^{fuz}: constraints of the form $\{< [\alpha_1], = [\alpha_2], > [\alpha_3]\}$;
  - PA_c^{fuz}: as PA^{fuz} but with the additional requirement that $\alpha_2 \geq \{\alpha_1, \alpha_3\}$;

- Fuzzy extensions of classical sub-algebras IA_p and IA_c (see Section 2.1.3):

  - IA_p^{fuz}: IA^{fuz} constraints that can be expressed as conjunctions of PA^{fuz} constraints on the intervals endpoints;
  - IA_c^{fuz}: IA^{fuz} constraints that can be expressed as conjunctions of PA_c^{fuz} constraints on the intervals endpoints.

In [5] it is shown that both IA_p^{fuz} and IA_c^{fuz} are tractable. In particular, the result for IA_c^{fuz} is obtained by noting that even in the case with preferences path consistency entails minimality (thus yielding a complexity of $O(kn^3)$), and the tractability of IA_p^{fuz} is obtained from enforcing minimality on 4-subnetworks (thus yielding a complexity of $O(kn^4)$). Also the Ord-Horn maximal tractable subclass [85] (see Section 2.1) has been extended with fuzzy preference [5] (denoted with H^{fuz}) by defining a IA^{fuz} relation to be in H^{fuz} if all its α-cuts are Ord-Horn relations. Moreover, fuzzyfication is shown to preserve maximality in terms of tractability.

Other approaches have used fuzzy sets and their theory to represent uncertainty, rather than preferences, in the context of Allen's temporal relations.

Along this line of research, in [32] fuzzy counterparts of the classical point relations are proposed. For example, in the case of points, "<" is interpreted as "much smaller" and "=" as "approximately equal". Given these counterparts, similar ones are derived for Allen's interval relations and for the operations performed on them.

The modeling of information about historical events motivates the model for using fuzzy sets to represent uncertainty and vagueness in qualitative temporal constraints presented in [84]. Here fuzzy values are used to express the confidence level that a given time belongs to an interval modeling an event. A similar approach is taken in [86] where fuzzy relations are allowed also on crisp intervals (as opposed to only on fuzzy intervals).

3.1.2   PREFERENCES IN QUANTITATIVE FRAMEWORKS

The extension of quantitative temporal models to preferences arose in the literature [55] as a special instance of the more general extension of hard constraints to soft constraints [11]. The main idea behind such an extension is to equip constraint problems with an algebraic structure that provides preference values and operators to manipulate them. Such a structure is called a c-semiring and is represented as a tuple $\langle A, +, \times, 0, 1 \rangle$ where A is a preference set, $+$ is an idempotent additive operator which induces the ordering over the preferences and $\times$ is an associative operator used to aggregate several preferences into a single one. In order for this algebraic structure to be a c-semiring certain properties have to hold on the operators (see [11] for further details). Different c-semirings yield a different preference semantics. For example, if the set of preferences is the interval between 0 and 1 (i.e., $A = [0, 1]$), the ordering is induced by the additive operator max, and preferences are aggregated via the multiplicative operator min. we obtain fuzzy preferences. If instead preferences are positive integers, and they are aggregated by the arithmetic sum and ordered with min then we obtain a modeling of costs.

3.1.3   SIMPLE TEMPORAL PROBLEMS WITH PREFERENCES (STPPS)

Although very expressive, STPs are able to model just *hard* temporal constraints. This means that all constraints have to be satisfied, and that the solutions of a constraint are all equally satisfying. However, in many real-life scenarios these two assumptions do not hold. In particular, sometimes some solutions are preferred with respect to others. Therefore the global problem is not to find a way to satisfy all constraints, but to find a way to satisfy them optimally, according to the preferences specified. On the other hand, some problems may be over-constrained and have no solutions. In such cases, the only alternative to not returning any solution may be that of relaxing some of the constraints. However, it may be desirable to encode in the relaxation the fact that solutions that are "closest" to satisfying the original constraints should be preferred.

To address these kind of problems a framework was introduced [55, 57] in which each temporal constraint is associated with a preference function specifying the preference for each distance or duration.

Definition 3.1   A *soft simple temporal constraint* is a 4-tuple $\langle (X, Y), I, A, f \rangle$ consisting of

- an ordered pair of variables (X, Y) over the integers, called the scope of the constraint;

- an interval $I = [a, b]$ where a and b are integers such that $a \leq b$;

- a set of preferences A;

- a preference function f, where $f : [a, b] \to A$, is a mapping of the elements belonging to interval I into preference values, taken from set A.

An assignment v_x and v_y to the variables X and Y is said to *satisfy* the constraint $\langle (X, Y), I, A, f \rangle$ iff $a \leq v_y - v_x$. In such a case, the preference associated to the assignment by the constraint is $f(v_y - v_x)$.

Definition 3.2   STPP.   Given a c-semiring $S = \langle A, +, \times_S, 0, 1 \rangle$, a *Simple Temporal Problem with Preferences* over S is a pair $\langle V, C \rangle$, where V is a set of variables and C is a set of soft simple temporal constraints over pairs of variables in V and with preferences in A.

Given an STPP $\langle V, C \rangle$ over a c-semiring S, a *solution t* is an assignment to all the variables in V that satisfies all the constraints in C. t is said to satisfy a constraint c in C with preference p if the projection of t over the pair of variables of c satisfies c with an associated preference equal to p, written $pref(t, c) = p$.

In an STPP each solution has a *global preference value*, obtained by combining, via the $\times$ operator of the c-semiring, the preference levels at which the solution satisfies the constraints in C. The optimal solutions of an STPP are those solutions which are not dominated by any other solution in terms of global preference.

For example, consider the c-semiring $S_{fuzzy} = \langle [0, 1], \max, \min, 0, 1 \rangle$, used for fuzzy constraint solving [100]. The global preference value of a solution will be the minimum of all the preference values associated with the distances selected by this solution in all constraints, and the best solutions will be those with the maximal value.

Another example is the c-semiring $S_{csp} = \langle \{ false, true \}, \vee, \wedge, false, true \rangle$, used to describe hard constraint problems [69]. Here there are only two preference values: *true* and *false*, the preference value of a complete solution will be determined by the logical *and* of all the local preferences, and the best solutions will be those with preference value *true* (since *true* is better than *false* in the order induced by using the logical or as the $+$ operator of the c-semiring). This c-semiring thus recasts the classical TCSP framework into STPPs by using preference functions which map into *true* elements belonging to allowed intervals and into *false* the forbidden regions lying in between.

STPPs are expressive enough to represent many real life scenarios. Figure 3.1 depicts an STPP that models the scenario in which there are three events to be scheduled on a satellite: the start time (**Ss**) and ending time (**Se**) of a slewing procedure and the (**Is**) starting time of an image retrieval. The slewing activity in this example can take from 3 to 10 units of time, but it is preferred that it takes the shortest time possible. The image taking can start any time between 3 and 20 units of time after the slewing has been initiated. The third constraint, on variables **Is** and **Se**, models the fact that it is better for the image taking to start as soon as the slewing has stopped.

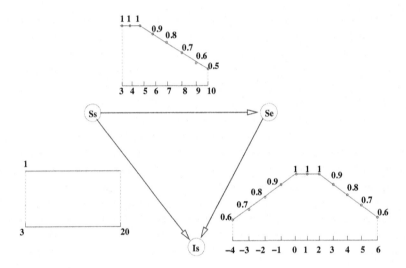

Figure 3.1: Example of an STPP.

We recall that STPs are polynomially solvable (see Section 2.2.1); thus one might speculate that the same is true for STPPs. However, it was shown in [55] that, in general, STPPs fall into the class of NP-hard problems. The results were obtained via a reduction from TCSPs, which are NP-hard problems (see Chapter 2), obtained, as briefly described above, by using preferences from the classical c-semiring $S_{csp} = \langle\{false, true\}, \vee, \wedge, false, true\rangle$. Tractability conditions have, however, been identified for STPPs as we discuss below.

The operations defined on temporal constraints can be extended to deal with preferences. The underlying idea is the same, that is, to induce new constraints that are as explicit as possible.

- Given two STPP constraints, $T_{ij}^1 = \langle I_{ij}^1, f_{ij}^1 \rangle$ and $T_{ij}^2 = \langle I_{ij}^2, f_{ij}^2 \rangle$, defined on the same variables X_i and X_j and a c-semiring S, we define the *intersection* $T_{ij}^1 \oplus_S T_{ij}^2$, as the soft temporal constraint $\langle I_{ij}^1 \oplus I_{ij}^2, f \rangle$, defined on the same variables, where:

– $I_{ij}^1 \oplus I_{ij}^2$ is the intersection of intervals I_{ij}^1 and I_{ij}^2, and

– $f(a) = f_{ij}^1(a) \times_S f_{ij}^2(a)$ for all $a \in I_{ij}^1 \oplus I_{ij}^2$.

• Given two STPP constraints, $T_{ij} = \langle I_{ik}, f_{kj} \rangle$ defined on variables X_i and X_k and $T_{kj} = \langle I_{kj}, f_{kj} \rangle$, defined on variables X_k and X_j and a c-semiring S, we define *composition* $T_{ik} \otimes_S T_{kj}$, as the soft constraint $T_{ij} = \langle I_{ij}, f_{ij} \rangle$ defined on varible X_i and X_j where:

– $r \in I_{ij}$ if and only if there exists a value $t_1 \in I_{ik}$ and $t_2 \in I_{kj}$ such that $r = t_1 + t_2$, and

– $f(a) = \sum\{ f_{ik}(a_1) \times_S f_{kj}(a_2) \mid a = a_1 + a_2, a_1 \in I_{ik}, a_2 \in I_{kj} \}$.

Examples of an intersection and composition of STPP constraints are shown in Figures 3.2 and 3.3.

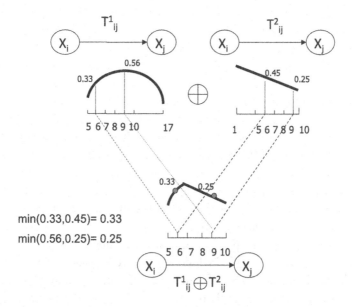

Figure 3.2: Intersection of two STPP constraints defined on variables X_i and X_j.

The *path-induced* constraint on variables X_i and X_j is $R_{ij}^{path} = \oplus_S \forall k (T_{ik} \otimes_S T_{kj})$, i.e., the result of performing $\oplus_S$ on each way of generating paths of length two from X_i to X_j. A constraint T_{ij} is *path-consistent* if and only if $T_{ij} \subseteq R_{ij}^{path}$, i.e., T_{ij} is at least as strict as R_{ij}^{path}. A STPP is path-consistent if and only if all its constraints are path-consistent.

In order to use path consistency for finding optimal solutions of STPPs, we must ensure that by enforcing it on any constraint of a given STPP, the resulting STPP is equivalent to the given one, that is, it has the same set of solutions with the same global preferences. The assumption that guarantees this is that the c-semiring must have an idempotent multiplicative operator [57];

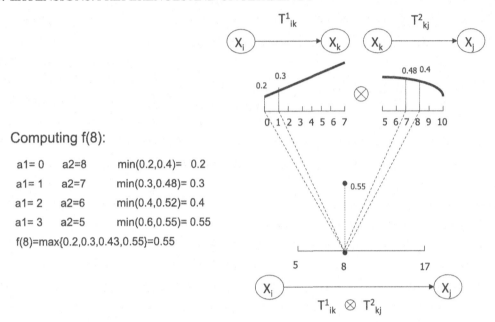

Computing f(8):

   a1= 0   a2=8   min(0.2,0.4)= 0.2

   a1= 1   a2=7   min(0.3,0.48)= 0.3

   a1= 2   a2=6   min(0.4,0.52)= 0.4

   a1= 3   a2=5   min(0.6,0.55)= 0.55

  f(8)=max{0.2,0.3,0.43,0.55}=0.55

Figure 3.3: Composition of two STPP constraints. The figure details how the preference value of an element, i.e., 8, of the interval of the resulting constraint is computed.

this being a sufficient condition can be derived directly from conditions which are sufficient for equivalence in generic soft CSPs (SCSPs) [11].

Under the same condition, applying the path consistency operation to a set of constraints (rather than just one) returns an STPP which is always the same, independently of the order of application [57]. It follows that an STPP can be transformed into an equivalent path-consistent Temporal CSP with Preferences by applying operation $T_{ij} := T_{ij} \oplus_S (T_{ik} \otimes_S T_{kj})$ to all constraints T_{ij} until no change occurs on any constraint. We will call this path consistency enforcing algorithm *STPP_PC-2*. As discussed in Chapter 2 enforcing path consistency is polynomial for STPs and applying it to an STN is sufficient to find a solution. Therefore, it is easy to infer that the hardness result for STPPs derives either from the nature of the c-semiring or from the shape of the preference functions.

When the preference functions are linear, and the c-semiring chosen is such that its two operations maintain such linearity when applied to the initial preference function, it can be seen that the initial STPP can be written as a linear programming problem, solving which is tractable [21]. To see this, consider any pair of variables X and Y, and an interval for the constraint over X and Y, say $[a, b]$, with associated linear preference function f. The information given by such intervals can be represented by the following inequalities and equation: $X - Y \leq b, Y - X \leq -a$

and $f_{X,Y} = c_1(X - Y) + c_2$, where c_1 and c_2 are appropriately defined constants. Then if we choose the fuzzy c-semiring $S_{FCSP} = \langle [0, 1], max, min, 0, 1 \rangle$, the global preference value V will satisfy the inequality $V \leq f_{X,Y}$ for each preference function $f_{X,Y}$ defined in the problem, and the objective is $max(V)$. If instead we choose the c-semiring $\langle R, min, +, \infty, 0 \rangle$, where the objective is to minimize the sum of the preference level, we have $V = f_1 + \cdots + f_n$[1] and the objective function is $min(V)$. In both cases, the resulting set of formulae constitutes a linear programming problem.

Linear preference functions are expressive enough for many cases, but there are also several situations in which we need preference functions which are not linear. A typical example arises when we want to state that the distance between two variables must be as close as possible to a single value. Then, unless this value is one of the extremes of the interval, the preference function is convex, but not linear. Another case is one in which preferred values are as close as possible to a single distance value, but in which there are some subintervals where all values have the same preference. In this case, the preference criteria define a *step function*, which is not even convex.

Let us now consider the class of semi-convex functions which includes linear, convex, and also some step functions. Semi-convex functions are such that, if one draws a horizontal line anywhere in the Cartesian plane corresponding to the graph of the function, the set of X such that $f(X)$ is not below the line forms a single interval. More formally, a *semi-convex* function f is one such that, for all $y \in \Re^+$, the set $F(X)\{x \in X$ such that $f(x) \geq y\}$ forms an interval. For example, the *close to k* criteria cannot be coded into a linear preference function, but it can be specified by a semi-convex preference function, which could be $f(x) = x$ for $x \leq k$ and $f(x) = 2k - x$ for $x > k$. Figure 3.4 shows some examples of semi-convex and non-semi-convex functions.

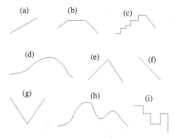

Figure 3.4: Examples of semi-convex functions [(a)–(f)] and non-semi-convex functions [(g)–(i)].

Semi-convex functions are closed under the operations of intersection and composition, when the c-semiring is totally ordered and the multiplicative operator is idempotent [57]. For example, this happens in the fuzzy c-semiring. It can also be shown that, given a path consistent STPP with semi-convex functions defined on a c-semiring with an idempotent multiplicative operator, then all its preference functions have the same maximum preference, say M. Moreover,

---

[1]In this context, the "+" is to be interpreted as arithmetic "+."

if we consider the STP obtained by considering on each constraint only the subinterval mapped by the preference function into M, then its set of solutions coincides with the set of optimal solutions of the original STPP.

This line of reasoning underlies one of the solvers which was proposed for STPPs [57], which we call *Path-solver*. *Path-solver* takes as input an STPP P with semi-convex preference functions, and returns an optimal solution of the given problem, by first applying path consistency, thus producing a new equivalent problem P'. Then, an STP corresponding to P' is constructed, by taking only the subintervals of P' corresponding to the best preference level and forgetting about the preference functions. Finally, a backtrack-free search is performed to find a solution of the STP (which is already in its minimal network form), specifically the earliest one is returned by the search procedure. All these steps are polynomial, so the overall complexity of solving an STPP with the above assumptions is polynomial, and, in particular is equal to $O(n^3 r l)$ for an STPP with n variables, maximum size of intervals r, and l distinct totally ordered preference levels.

In Figure 3.5 we show the effect of applying *Path-Solver* on the example depicted in Figure 3.1. As it can be seen, the interval on the constraint on variables **Ss** and **Is** has been reduced from [3,20] to [3,14] and some preferences on all the constraints have been lowered. It is easy to see that the optimal preference of the STPP is 1 and the minimal STP containing all optimal solutions restricts the duration of the slewing to interval [4,5], the interleaving time between the slewing start and the image start to [3,5] and the interleaving time between the slewing stop and the image start to [0,1].

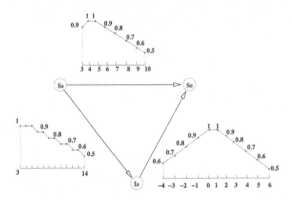

Figure 3.5: The STPP representing the scenario depicted in Figure 3.1 after applying STPP_PC-2.

In [57] another algorithm for finding an optimal solution of fuzzy STPPs, called *Chop-Solver* was proposed. The idea behind *Chop-Solver* is similar to the concept of α-cut in Fuzzy Set theory [119]. In particular, given an STPP and underlying preference set A, let $y \in A$ and $\langle I, f \rangle$ be a soft constraint defined on variables X_i and X_j in the STPP, where f is semi-convex. Consider the interval defined by $\{x : x \in I \cap f(x) \geq y\}$ (because f is semi-convex this set defines a single interval for any choice of y). In fuzzy set theory, $\{x : x \in I \cap f(x) \geq y\}$ is called the y-level cut of function f. Let this interval define a constraint on the same pair X_i and X_j. Performing this transformation on each soft constraint in the original STPP results in an STP, which we refer to as STP_y. Let opt be the highest preference value (in the ordering induced by the c-semiring) such that STP_{opt} has a solution. It was shown that the solutions of STP_{opt} are the optimal solutions of the given STPP. This result implies that finding an optimal solution of the given STPP with semi-convex functions reduces to a two step search process, consisting of iteratively choosing a cut level w and then solving the corresponding problem STP_w, until STP_{opt} is found. Both phases can be performed in polynomial time, and hence the entire process is tractable. This approach allows us to relate the tractability of STPPs with semi-convex functions with the polynomial class induced by the LP mentioned earlier. Such mapping can be done by considering each STP_w as having constant preference functions equal to w. Thus, solving the STPP reduces to solving a finite number of linear programs. Since semi-convexity is preserved only if the multiplicative operator of the c-semiring is idempotent then only S_{fuzzy} can be considered as the underlying c-semiring. *Chop-Solver*, in fact, works with STPPs with semi-convex functions based on the fuzzy c-semiring. This means that the set of preferences we are considering are contained in the interval $[0, 1]$. The solver finds an optimal solution of the STPP identifying first STP_{opt} and then returning its earliest or latest solution. Preference level opt is found by performing a binary search among the preference levels in $[0, 1]$. The bound on the precision of a preference, that is the maximum number of decimal coded digits, implies that the number of search steps is always finite. In particular, the complexity of *Chop-Solver* can be defined as $O(precision \times n^3)$ where n is the maximum number of steps allowed in the binary search.

For example *Chop-Solver* when applied to the triangular STPP shown in Figure 3.1 will stop the search when it reaches preference level 1 and finds that the STP obtained chopping the STPP at 1 is consistent and has as its minimal network the same found by Path-Solver: [4,5] on the constraint on **Ss** and **Se**, [3,5] on the constraint on **Ss** and **Is**, and [0,1] on the constraint **Is** and **Se**.

Chop-Solver was shown experimentally to significantly outperform *Path-Solver* in terms of time. This is mainly due to the discretization of the preference functions required to perform the operations involved in path consistency. It should be noted, however, that *Chop-Solver* requires a fast computation of the cut of the preference functions in order to perform well.

3.1.4   OTHER EXTENSIONS

Additional work has followed the introduction of preferences in STPs. For example, it has been shown that preferences can be embedded also in inherently intractable problems such as those with disjunctions. In particular, in [88] disjunctive temporal problems (DTPs) (see Chapter 2.2.3) were augmented with fuzzy preferences. The solving techniques adopted there follow a cut-based approach, much like the one adopted in *Chop-Solver*, requiring however more sophisticated projection techniques and a solver for DTPs. Moreover, it was shown that the significant increase in expressive power which preferences bring is computationally inexpensive: the complexity class of the problems does not change by adding preferences and the added cost turns out to be worst case polynomial in the number of time points.

As we have discussed above, it appears clear that using fuzzy preferences is crucial in terms of taming complexity. However, this does come at a price. Fuzzy preferences model settings in which one wants to maximize the minimally preferred local preference. This is sometimes called the *maximin* approach and can be suitable in scenarios where a pessimistic attitude focussing on the worst violation entailed by a solution is appropriate. Given a solution, the weakest links with respect to that solution are those constraints on which the preference assigned to the projection of the solution is the same as the global preference of that solution. The maximin criterion offers what amounts to a coarse method for comparing solutions, based on the minimal preference value over all the projections of the solutions to local preference functions. This is known as the "drowning effect". Consequently, following this criterion may be insufficient to find solutions that are intuitively more globally preferable. In [56] this problem is tackled by designing an algorithm that finds solutions which are optimal according to the Pareto criterion. A solution of a fuzzy STPP is Pareto optimal if there is no other solution that gets higher or equal preferences on the projections of all constraints and does strictly better on at least one constraint. The solution proposed in [56] is based on the intuition that if a constraint solver using the maximin criterion could iteratively "ignore" the weakest link values (i.e., the values that contributed to the global solution evaluation) then it could eventually recognize solutions that dominate others in the Pareto sense. This intuition is formalized by a procedure wherein the original STPP is transformed by iteratively selecting a weakest link w.r.t. the set of optimal solutions, changing the constraint in such a way that it can effectively be "ignored," and solving the transformed problem. The concept of "ignoring" weakest link values is implemented by a two-step process of restricting the weakest links to their intervals of optimal temporal values, while eliminating their restraining influence by resetting their preferences to a single, "best" value (1 in the fuzzy context). It is easy to see that the complexity of this procedure is in the worst case equal to that of applying *Chop-Solver* as many times as the number of constraints which is $O(n^2)$ if n represents the number of variables in the STPP.

The procedure described above does not however return any Pareto solution but rather solutions optimizing a stronger optimality criterion than Pareto which is defined in [75] as "stratified egalitarianism". This notion of optimality is inspired by an economic metaphor for which, given a fixed "poverty" threshold a solution S' has a better overall quality than another solution S if some members below the poverty line in S are better off in S', even if some of those above the poverty line in S are made worse off in S' (as long as they do not drop below the poverty line).

A natural, and more robust alternative evaluates solutions by summing the preference values, and ordering them based on preferences toward larger values. This criterion, which also ensures Pareto optimality since every maximum sum solution is Pareto optimal, is called "utilitarian." The main drawback to this alternative is that the ability to solve STPPs tractably is no longer apparent. The reason is that the formalization of utilitarianism as a c-semiring forces the multiplicative operator (in this case, sum), not to be idempotent (i.e., $a + a \neq a$), a condition required in the proof that a local consistency approach is applicable to the soft constraint reasoning problem. However, in [75] it has been shown that finding utilitarian optimal solutions of an STPP can be done in polynomial time if the preference functions are piecewise linear and convex. Piecewise linear preference functions characterize soft constraints in many real scheduling problems; for example, in vehicle routing (where the best solutions are close to desired start times) and in dynamic rescheduling (where the goal is to find solutions that minimally perturb the original schedule).

This line of research addressing utilitarian optimal solutions was further pursued in [89] where the authors present a greedy and a complete algorithm for solving this problem for STPPs without restrictions on the preference functions. Although relaxing this assumption implies moving to a non-tractable class of problems, the experimental results in [89] show how good quality solutions can be obtained in a few iterations of the greedy procedure, how near-optimal solutions can be found in a number of iterations bounded by the square of the number of constraints, and how the complete algorithm outperforms a general branch and bound approach and exhibits compelling anytime properties.

3.2 UNCERTAINTY

3.2.1 SIMPLE TEMPORAL PROBLEMS WITH UNCERTAINTY

When reasoning concerns activities that an agent performs interacting with an external world, uncertainty is often unavoidable. TCSPs (see Chapter 2) assume that all activities have durations under the control of the agent. Simple Temporal Problems with Uncertainty (STPUs) [115] extend STPs by distinguishing contingent events, whose occurrence is controlled by exogenous factors often referred to as "Nature".

As in STPs, activity durations in STPUs are modelled by intervals. The start times of all activities are assumed to be controlled by the agent (this brings no loss of generality). The end times, however, fall into two classes: requirement (called free in [115]) and contingent. The former, as in STPs, are decided by the agent, but the agent has no control over the latter: it only can observe their occurrence after the event; observation is supposed to be known immediately after the event. The only information known prior to observation of a time-point is that nature will respect the interval on the duration. Durations of contingent links are assumed to be independent.

In an STPU, the variables are thus divided into two sets depending on the type of time-points they represent:

- executable time-points: variables, b_i, whose time is assigned by the executing agent;

- contingent time-points: variables, e_i, whose time is assigned by the external world.

The distinction on variables leads to constraints which are also divided into two sets, requirement and contingent, depending on the type of variables they constrain. Note that as in STPs all the constraints are binary. Formally:

- a requirement constraint (or link) r_{ij}, on generic time-points t_i and t_j,[2] is an interval $I_{ij} = [l_{ij}, u_{ij}]$ such that $l_{ij} \leq \gamma(t_j) - \gamma(t_i) \leq u_{ij}$ where $\gamma(t_i)$ is a value assigned to variable t_i

- a contingent constraint g_{hk}, on executable point b_h and contingent point e_k, is an interval $I_{hk} = [l_{ij}, u_{ij}]$ which contains all the possible durations of the contingent event represented by b_h and e_k.

The formal definition of an STPU is the following:

Definition 3.3 STPU. A *Simple Temporal Problem with Uncertainty* is a 4-tuple $N = \{X_e, X_c, R_r, R_c\}$ such that:

- $X_e = \{b_1, \ldots, b_{n_e}\}$: is the set of executable time-points;

- $X_c = \{e_1, \ldots, e_{n_c}\}$: is the set of contingent time-points;

- $R_r = \{c_{i_1 j_1}, \ldots, c_{i_C j_C}\}$: is the set of requirement constraints;

- $R_c = \{g_{i_1 j_1}, \ldots, g_{i_G j_G}\}$: is the set of contingent constraints.

Example 3.4 This is an example taken from [115], which describes a scenario which can be modeled using an STPU. Consider two activities *Cooking* and *Having dinner*. Assume you don't want to eat your dinner cold. Also, assume you can control when you start cooking and when the dinner starts but not when you finish cooking or when the dinner will be over. The STPU modeling this example is depicted in Figure 3.6. There are two executable time-points {*Start-cooking, Start-dinner*} and two contingent time-points {*End-cooking, End-dinner*}. Moreover, the contingent constraint on variables {*Start-cooking, End-cooking*} models the uncontrollable duration of fixing dinner which can take anywhere from 20 to 40 minutes; the contingent constraint on variables {*Start-dinner, End-dinner*} models the uncontrollable duration of the dinner that can last from 30 to 60 minutes. Finally, there is a requirement constraint on variables {*End-cooking, Start-dinner*} that simply bounds to 10 minutes the time between when the food is ready and when the dinner starts.

A *control sequence* δ is an assignment to executable time-points. It is *partial* if it assigns values to a proper subset of the executables; otherwise, it is *complete*.

A *situation* ω is a set of durations on contingent constraints. Situations can be partial or complete as well.

---

[2]Note that t_i and t_j can be either contingent or executable time-points.

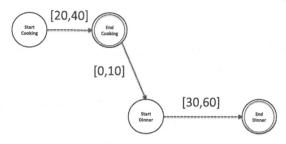

Figure 3.6: The STPU for Example 3.4. Contingent time-points are double-circled; contingent constraints are represented with a dashed arrow.

A *schedule* is a complete assignment to all the time-points in X_e and X_c. A schedule T identifies a control sequence, δ_T, consisting of all the assignments to the executable time-points, and a situation, ω_T, which is the set of all the durations identified by the assignments in T on the contingent constraints. $Sol(P)$ denotes the set of all schedules of an STPU.

It is easy to see that each situation corresponds to an STP. In fact, once the durations of the contingent constraints are fixed, there is no more uncertainty in the problem, which becomes an STP, called the *underlying STP*.

A *projection* P_ω, corresponding to a situation ω, is the STP obtained by leaving all requirement constraints unchanged and replacing each contingent constraint g_{hk} with the constraint $\langle[\omega_{hk}, \omega_{hk}]\rangle$, where ω_{hk} is the duration of event represented by g_{hk} in ω. $Proj(P)$ is the set of all projections of an STPU P.

3.2.2 CONTROLLABILITY

It is clear that in order to solve a problem with uncertainty all possible situations must be considered. The notion of consistency defined for STPs does not apply since it requires the existence of a single schedule, which is not sufficient for STPUs where all situations are equally possible.[3] For this reason, in [115], the notion of controllability has been introduced. *Controllability* of an STPU is, in some sense, the analogue of consistency of an STP. Controllable means the agent has a means to execute the time-points under its control, subject to all constraints. The notion of controllability is expressed in terms of the ability of the agent to find, given a situation, an appropriate control sequence. This ability is identified with having a *strategy*, that is, a mapping $S : Proj(P) \rightarrow Sol(P)$, such that for every projection P_ω, $S(P_\omega)$ is a schedule which induces the durations in ω on the contingent constraints. Further, a strategy is said to be *viable* if, for every projection P_ω, $S(P_\omega)$ is a solution of P_ω.

---

[3]In [110] STPUs have been augmented to include probability distributions over the possible situations.

We will write $[S(P_\omega)]_x$ to indicate the value assigned to executable time-point x in schedule $S(P_\omega)$, and $[S(P_\omega)]_{<x}$ the *history* of x in $S(P_\omega)$, that is, the set of durations of contingent constraints which occurred in $S(P_\omega)$ before the execution of x, i.e., the partial solution so far.

In [115], three notions of controllability are introduced for STPUs.

Strong Controllability

The first notion is, as the name suggests, the most restrictive in terms of the requirements that the control sequence must satisfy. An STPU P is *Strongly Controllable* (SC) if and only if there is an execution strategy S s.t. $\forall P_\omega \in Proj(P)$, $S(P_\omega)$ is a solution of P_ω, and $[S(P_1)]_x = [S(P_2)]_x$, $\forall P_1, P_2$ projections and for every executable time-point x.

In words, an STPU is *strongly controllable* if there is a fixed execution strategy that works in all situations. This means that there is a fixed control sequence that will be consistent with any possible scenario of the world. Thus, the notion of strong controllability is related to that of conformant planning, i.e., planning for every contingency, and is therefore a very strong requirement. As [115] suggest, SC may be relevant in some applications where the situation is not observable at all or where the complete control sequence must be known beforehand (for example in cases in which other activities depend on the control sequence, as in the production planning area).

In [115] a polynomial time algorithm for checking if an STPU is strongly controllable is proposed. The main idea is to rewrite the STPU given in input as an equivalent STP only on the executable variables. What is important to notice is that this algorithm takes as input an STPU $P = \{X_e, X_c, R_r, R_c\}$ and returns in output an STP defined on variables X_e. The STPU in input is strongly controllable if and only if the derived STP is consistent. Moreover, every solution of the STP is a control sequence which guarantees strong controllability for the STPU. When the STP is consistent, the output of STPU is its minimal form.

In [115] it is shown that the complexity of this procedure is $O(n^3)$, where n is the number of variables.

Weak Controllability

The notion of controllability with the fewest restrictions on the control sequences is called Weak Controllability. An STPU P is *Weakly Controllable* (WC) if and only if $\forall P_\omega \in Proj(P)$ there is a strategy S_ω s.t. $S_\omega(P_\omega)$ is a solution of P_ω.

In words, an STPU is *weakly controllable* if there is a viable global execution strategy: there exists at least one schedule for every situation. This can be seen as a minimum requirement since, if this property does not hold, then there are some situations such that there is no way to execute the controllable events in a consistent way. It also looks attractive since, once an STPU is shown to WC, as soon as one knows the situation, one can pick out and apply the control sequence that matches that situation. Unfortunately in [115] it is shown that this property is not so useful in classical planning. Nonetheless, WC may be relevant in specific applications (as large-scale warehouse scheduling) where the actual situation will be totally observable before (possibly *just before*) the execution starts, but one wants to know in advance that, whatever the situation, there will always be at least one feasible control sequence.

In [115] it is conjectured and in [78] it is proven that the complexity of checking weak controllability is co-NP-hard. The algorithm proposed for testing WC in [115] is based on a classical enumerative process and a lookahead technique.

Strong Controllability implies Weak Controllability [115]. Moreover, an STPU can be seen as an STP if the uncertainty is ignored. If enforcing path consistency removes some elements from the contingent intervals, then these elements belong to no solution. If so, it is possible to conclude that the STPU is not weakly controllable.

Another useful, although more "operational" notion of controllability is *pseudo-controllability*. An STPU is *pseudo-controllable* if applying path consistency leaves the intervals on the contingent constraints unchanged.

Unfortunately, if path consistency leaves the contingent intervals untouched, we cannot conclude that the STPU is weakly controllable. That is, WC implies pseudo-controllability but the converse is false. In fact, weak controllability requires that given any possible combination of durations of all contingent constraints the STP corresponding to that projection must be consistent. Pseudo-controllability, instead, only guarantees that for each possible duration on a contingent constraint there is at least one projection that contains such a duration and it is a consistent STP.

Dynamic Controllability

In dynamic applications domains, such as planning, the situation is observed over time. Thus decisions have to be made even if the situation remains partially unknown. Indeed the distinction between Strong and Dynamic Controllability is equivalent to that between conformant and conditional planning. The final notion of controllability defined in [115] addresses this case. Here we give the definition provided in [76] which is equivalent but more compact.

Definition 3.5 Dynamic Controllability. An STPU P is *Dynamically Controllable* (DC) if and only if there is a strategy S such that $\forall P_1, P_2$ in $Proj(P)$ and for any executable time-point x:

1. if $[S(P_1)]_{<x} = [S(P_2)]_{<x}$ then $[S(P_1)]_x = [S(P_2)]_x$;

2. $S(P_1)$ is a solution of P_1 and $S(P_2)$ is a solution of P_2.

In words, an STPU is dynamically controllable if there exists a viable strategy that can be built, step-by-step, depending only on the observed events at each step. SC $\implies$ (implies) DC and DC $\implies$ WC. Dynamic Controllability, seen as the most useful controllability notion in practice, is also the one that requires the most complicated algorithm. Surprisingly, [76] and [79] proved that DC is polynomial in the size of the STPU representation.

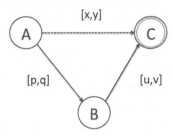

Figure 3.7: Contingent time-points are double-circled, contingent constraints are represented with a dashed arrow.

The algorithm proposed there is based on some considerations on triangles of constraints. The triangle shown in Figure 3.7 is a triangular STPU with one contingent constraint, AC, two executable time-points, A and B, and a contingent time-point C. Based on the sign of u and v, three different cases can occur:

- *Follow case ($v < 0$):* B will always follow C. If the STPU is path consistent then it is also DC since, given the time at which C occurs after A, by definition of path consistency, it is always possible to find a consistent value for B.

- *Precede case ($u \geq 0$):* B will always precede or happen simultaneously with C. Then the STPU is dynamically controllable if $y - v \leq x - u$, and the interval $[p, q]$ on AB should be replaced by interval $[y - v, x - u]$, that is by the sub-interval containing all the elements of $[p, q]$ that are consistent with each element of $[x, y]$.

- *Unordered case ($u < 0$ and $v \geq 0$):* B can either follow or precede C. To ensure dynamic controllability, B must wait either for C to occur first, or for $t = y - v$ units of time to go by after A. In other words, either C occurs and B can be executed at the first value consistent with C's time, or B can safely be executed t units of time after A's execution. This can be described by an additional constraint which is expressed as a *wait* on AB and is written $< C, t >$, where $t = y - v$. Of course if $x \geq y - v$ then we can raise the lower bound of AB, p, to $y - v$ (*Unconditional Unordered Reduction*), and in any case we can raise it to x if $x > p$ (*General Unordered reduction*) .

It can be shown that waits can be propagated (in [76] the term "regressed" is used) from one constraint to another: a wait on AB induces a wait on another constraint involving A, e.g., AD, depending on the type of constraint DB. In particular, there are two possible ways in which the waits can be regressed.

- *Regression 1:* assume that the AB constraint has a wait $\langle C, t \rangle$. Then, if there is any DB constraint (including AB itself) with an upper bound, w, it is possible to deduce a wait $\langle C, t - w \rangle$ on AD. Figure 3.8(a) shows this type of regression.

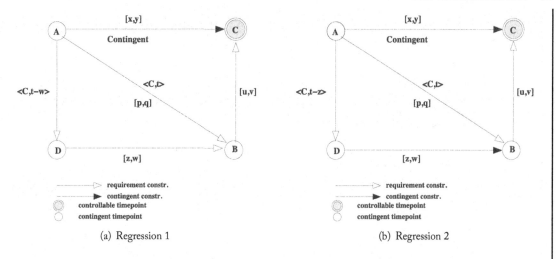

Figure 3.8: Regressions in Dynamic Controllability.

- *Regression 2:* assume that the AB constraint has a wait $\langle C, t \rangle$, where $t \geq 0$. Then, if there is a contingent constraint DB with a lower bound, z, and such that $B \neq C$, it is possible to deduce a wait $\langle C, t - z \rangle$ on AD. Figure 3.8(b) shows this type of regression.

The DC algorithm applies these rules to all triangles in the STPU and regresses all possible waits. If no inconsistency is found, that is, no requirement interval becomes empty and no contingent interval is squeezed, the STPU is DC and the algorithm returns an STPU where some constraints may have waits to satisfy, and the intervals contain elements that appear in at least one possible dynamic strategy. This STPU can then be given to an execution algorithm which dynamically assigns values to the executables according to the current situation.

The execution algorithm observes, as the time goes by, the occurrence of the contingent events and accordingly executes the controllables. For any controllable B, its execution is triggered if it is (1) *live*, that is, if current time is within its bounds, it is (2) *enabled*, that is, all the executables constrained to happen before have occurred, and (3) all the waits imposed by the contingent timepoints on B have expired.

The complexity of the DC algorithm is $O(n^3 r)$, where n is the number of variables and r is the number of elements in an interval. Since the polynomial complexity relies on the assumption of a bounded maximum interval size, the result in [76] allows only to conclude that DC is *pseudo-polynomial*. A DC algorithm of "strong" polynomial complexity is presented in [79]. The new algorithm differs from the previous one mainly because it manipulates the distance graph rather than the constraint graph of the STPU. In particular, given an STPU $N = \{X_e, X_c, R_r, R_c\}$, a graph (X, E) is associated with N called the *labeled distance graph* and defined as follows:

- $X = X_e \cup X_c$, that is, there is a node for each executable and contingent variable in N;

- for each requirement constraint in R_r of type $X \xrightarrow{[u,v]} Y$, there are two *ordinary* edges in E: $X \xrightarrow{v} Y$ and $Y \xrightarrow{-u} X$, representing, respectively, $Y - X \leq v$ and $X - Y \leq -u$;

- for each contingent constraint in R_c $A \xrightarrow{[x,y]} C$, there is a *lower-case* edge in E: $A \xrightarrow{c:x} C$ representing the fact that $C - A$ may get its minimum value x, and an *upper-case* edge $C \xrightarrow{-y} A$, representing the fact that $C - A$ may get its minimum value y.

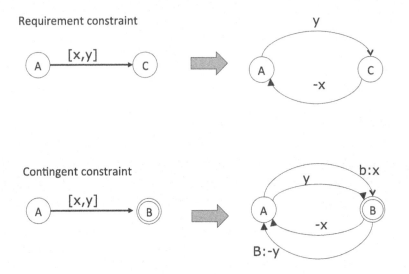

Figure 3.9: Graphical depiction of the correspondence between STNU constraints and constraints in the labeled distance graph. Note that "B" and "b" stand for the name of the contingent variable involved. Lower/upper case edges have the value of the interval corresponding, respectively, to the earliest/latest occurrence of B.

Figure 3.9 depicts the transformation of STNU constraints into constraints of the labeled graph. The algorithm in [79] redefines the original inference rules in terms of this new structure and works by recursively generating new edges. Its complexity is $O(n^5)$. In [77] the author further improves on this worst case complexity by presenting a DC-checking algorithm that runs in $O(n^4)$ obtained by targeting a specific type of edges. However, Morris in [77] conjectures there is an additional processing of $O(n^4)$ is necessary to prepare the network for execution. This conjecture is disproved in [48] where Hunsberger proves that execution can be launched directly on the network given in output by Morris's algorithm via an incremental, real-time algorithm that spreads out the (n^4)-computations over the entire execution time.

In [47] a problem with the original definition of the strategy in [76] was identified and fixed. In more detail, it was observed that a dynamic execution strategy, as defined in [76], does not fully capture the prohibition against the knowledge of future events which is crucial for the notion of dynamic controllability. The problem is fixed in [47] by strengthening the definition of pre-history and strategy in the following way. Histories (called prehistories in [47]) are given in terms of a real number k, representing the time assigned by a schedule to a variable. In particular, $[S(P_\omega)]_{<k}$ the *history* relative to number k in $S(P_\omega)$, is defined as the set of durations of contingent constraints in $S(P_\omega)$ that finish before k. Thus, a strategy ensuring dynamic controllability is such that $\forall P_1, P_2$ in *Proj(P)* and for any executable time-point x, if $[S(P_1)]_x = k$ and $[S(P_1)]_{<x} = [S(P_2)]_{<x}$ then $[S(P_2)]_x = k$. Given this new definition, no changes are required to the DC algorithm which can be shown to be sound with small changes to the original proof.

In [47] the author also introduced a new and more practical definition of dynamic execution strategies. The new characterization is based directly on the kinds of real-time execution decisions that a planning agent must make, namely, to either *"wait until something happens"* or *"if nothing happens before a time t, then to execute the time-points in a given set"*. Such strategies can be derived using incremental computations.

3.2.3   CONDITIONAL TEMPORAL PROBLEMS

CTPs [111] extend temporal constraint satisfaction problems [26] by adding observation variables and by conditioning the occurrence of some events on the presence of some properties of the environment.

Definition 3.6   CTP.   A *Conditional Temporal Problem* is a tuple $< V, E, L, OV, O, \mathcal{P} >$ where:

- $\mathcal{P}$ is a set of Boolean atomic propositions;

- V is a set of variables;

- E is a set of temporal constraints between pairs of variables in V;

- $L : V \rightarrow Q^*$ is a function attaching conjunctions of literals in $Q = \{p_i : p_i \in P\} \cup \{\neg p_i : p_i \in P\}$ to each variable in V;

- $OV \subseteq V$ is the set of observation variables;

- $O : \mathcal{P} \rightarrow OV$ is a bijective function that associates an observation variable to a proposition.

Notice that the constraints in E may be of any type, that is, as in STPs, TCSPs, or DTPs. Most of the work present in the literature, however, assumes an underlying STP. The observation variable $O(A)$ provides the truth value for A. In V there is usually a variable denoting the "origin of time" time, set to 0. We will denote this variable by x_0. Thus, in CTPs, variables are labeled with conjunctions of literals, and the truth values of such labels are used to determine whether variables represent events that are part of the temporal problem. In a CTP, for a variable to be executed, its associated label must be true. The truth values of the propositions appearing in the labels are provided when the corresponding observation variables are executed.

The constraint graph of a CTP is a graph where nodes correspond to variables and edges to constraints. A node v is labeled with $L(v)$ and an edge c is labeled with the interval of constraint c. Labels equal to *true* (i.e., with no constraints defined for them) are not specified.

An *execution scenario s* is a conjunction of literals that partitions the set of variables in two subsets: the subset of the variables that will be executed because their label is true given s, and the subset of the other variables, that will not be executed. SC is the set of all scenarios. Given a scenario s, its *projection*, $Pr(s)$, is the set of variables that are executed under s and all the constraints between pairs of them. $Pr(s)$ is a non-conditional temporal problem.

Figure 3.10 shows an example inspired from [111]. The example is about a plan to go skiing at station $Sk1$ or $Sk2$, depending on the condition of road R. Station $Sk2$ can be reached in any case, while station $Sk1$ can be reached only if road R is accessible. If $Sk1$ is reachable, we choose to go there. Moreover, temporal constraints between X_0, $WSk1_e$, and $WSk2_e$ limit the arrival times at the skiing stations. In particular, station $Sk1$ should be reached before 11:00 am in order to beat the crowd while at station $Sk2$ a special discount is applied after 1:00 pm. The condition of road R can be assessed when arriving at village W. In the figure, variables XY_s and XY_e represent the start and the end time for the trip from X to Y. Node $O(A)$, where A = "road R is accessible" is HW_e. There are two scenarios, A on variables $\{x_0, HW_s, HW_e, WSk1_s, WSk1_e\}$ and $\neg A$ on variables $\{x_0, HW_s, HW_e, WSk2_s, WSk2_e\}$.

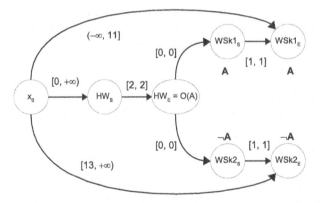

Figure 3.10: Example of Conditional Temporal Problem.

In CTPs there are three different notions of consistency depending on the assumptions made about the availability of observation information:

- *Strong Consistency (SC)*. Strong consistency applies when no information is available. A CTP is strongly consistent if there is a fixed way to assign values to all the variables so that all constraints are satisfied independently of the observations. A CTP is strongly consistent if and only if its non-conditional counterpart is consistent. Therefore, an algorithm to check SC of a CTP takes the same time as checking the consistency of an STP, which is polynomial.

- *Weak Consistency (WC).* Weak consistency applies when all information is available before execution. A CTP is weakly consistent if the projection of any scenario is consistent. Checking WC is a co-NP complete problem [111]. A brute force algorithm to check WC can test the consistency of all projections, possibly exploiting equivalent scenarios and shared paths.

- *Dynamic Consistency (DC).* Dynamic consistency (DC) assumes that information about observations becomes known during execution. A CTP is dynamically consistent if it can be executed so that the current partial solution can be consistently extended independently of the upcoming observations. In [111] it was shown how to convert the problem of dynamic consistency of a CTP to the consistency of an associated DTP. This makes the use off-the-shelf DTPs solvers sufficient for checking dynamic consistency (in exponential time).

The CTP depicted in Figure 3.10 is not DC. In fact, if A is true then we have to leave home before 8, since we want to get to $Sk1$ before 11 (($X_0, WSk1_e$) constraint) and it takes 2 hours to get to village W ((HW_s, HW_e) constraint) and an additional hour from W to $Sk1$ (($WSk1_s, WSk1_e$) constraint). If A is false, instead, we need to leave home after 10, since we want to be there after 1PM and from W it will take us an hour to get to $Sk2$ (($WSk1_s, WSk1_e$) constraint). However, being at village W is a precondition for the observation of proposition A and this fact prevents us from observing A before leaving home. Therefore we cannot distinguish between the two scenarios A and $\neg A$ in time to schedule our departure from home accordingly.

CTPs use logical formulas to define which nodes in the temporal network are active. The idea of active and inactive nodes is similar to Temporal Networks with Alternatives (TNA) [8] (see Section 2.2.4). The major difference is that a TNA uses implicit constraints to define which nodes are active via two types of branching constraints—alternative and parallel branching. In the parallel branching, the nodes in the branching are either all active or all inactive. For the alternative branching, either all nodes in the branching are inactive or only the root node and one of the rest nodes are active while the other nodes are inactive.

3.2.4 CONDITIONAL SIMPLE TEMPORAL NETWORKS WITH UNCERTAINTY

Recently [92] Conditional Simple Temporal Networks with Uncertainty (CSTNUs) have been proposed as a unifying class of problems combining observation nodes and branching for CTPs with contingent constraints from STPUs.[4] The main motivation behind this new framework is to accomodate workflow systems which are a popular tool for modeling business, manufacturing, and medical treatment processes. The key feature that enables merging CTPs with STNUs is the

---

[4]Notice that in the literature the STPU and STNU (for Simple Temporal Network with Uncertainty) acronyms are used interchangeably.

labeling of both nodes and edges with conjunctions of propositional literals the truth of which controls their actual existence. More formally, an CSTNU is defined as follows:

Definition 3.7 CSTNU.   A *Conditional STN with Uncertainty* is a tuple $\langle T, C, L, OT, O, P, L \rangle$ where:

- T is a set of real-valued time-points;

- P is a finite set of propositional letters;

- $L : T \to P^*$ maps a label to each time point;

- $OT \subseteq T$ is a set of observation time-points;

- $O : P \to OT$ is a bijection associating each observation time-point to a propositional letter;

- L is a set of contingent constraints;

- C is a set of labeled ST constraints of the form $(Y - X \leq \delta, l)$, where $X, Y \in T$, δ is a real number and $l \in P^*$;

- for any $(Y - X \leq \delta, l) \in C$, label l is satisfiable and subsumes both $L(X)$ and $L(Y)$;

- for any $p \in P$ and $t \in T$ if p or $\neg p$ appears in t's label then $L(t)$ subsumes $L(O(p))$ and $(O(p) - t \leq -\epsilon, L(t)) \in C$ for some $\epsilon > 0$;

- for each $(Y - X \leq \delta, l) \in C$ and each $p \in P$, if p or $\neg p$ appears in l, then l subsumes $L(O(p))$;

- $(T, \lfloor C \rfloor, L)$ is a STNU where $\lfloor C \rfloor = \{(Y - X \leq \delta) \mid (Y - X \leq \delta, l) \in C, \exists l\}$.

The graph of an CSTNU is just like the one of an STNU except that some nodes may be observation nodes and both nodes and edges are labeled. Instead, for CTPs, a *scenario* specifies the truth value to every propositional letter thus identifying univocally an STNU which is called a *projection* of the CSTNU. Similarly to STNUs, a *situation* ω specifies fixed durations for all the contingent links. These two notions are combined in CSTNUs into a *drama* (s, ω) in which all the uncertainty is resolved, resulting in an STP. Accordingly, an *execution strategy* σ maps dramas into schedules, that is, assignments of execution times to all the time-points. A *dynamic* execution strategy assignes executable time-points only based on *past* observations. An CSTNU is said to be dynamically controllable if it has at least one dynamic execution strategy which guarantees a solution for any possible drama. In [93] an algorithm for checking the dynamic controllability of a CSTNU is presented which extends the one for DC of STNUs in [77] to deal with propositional labels. The other core features of the algorithm are handling observation nodes and conjoining labels to enable propagation.

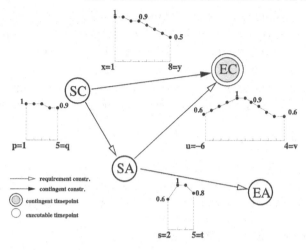

Figure 3.11: Example STPPU from the Earth Observing Satellites domain.

3.3   COMBINING PREFERENCES AND UNCERTAINTY

3.3.1   SIMPLE TEMPORAL PROBLEMS WITH PREFERENCES AND UNCERTAINTY

In [97] STPUs are extended to handle preferences by replacing ST constraints with soft temporal constraints. Thus an STPPU is a tuple $\langle N_e, N_c, L_r, L_c \rangle$ where N_e is the set of executable time-points, N_c is the set of contingent timepoints, L_r is a set of soft requirement constraints, and L_c is a set of soft contingent constraints.

Example 3.8   Consider as an example the following scenario from the Earth Observing Satellites domain [40]. Suppose a request for observing a region of interest has been received and accepted. To collect the data, the instrument must be aimed at the target before images can be taken. It might be, however, that for a certain period during the time window allocated for this observation, the region of interest is covered by clouds. The earlier the cloud coverage ends the better, since it will maximize both the quality and the quantity of retrieved data; but coverage is not controllable.

Suppose the time window reserved for an observation is from 1 to 8 units of time and that we start counting time when the cloud occlusion on the region of interest is observable. Also, suppose, in order for the observation to succeed, the aiming procedure must start before 5 units after the starting time, ideally before 3 units, and it actually can only begin after at least 1 time unit after the weather becomes observable. Ideally the aiming procedure should start slightly after the cloud coverage will end. If it starts too early, then, since the instrument is activated immediately after it is aimed, clouds might still occlude the region and the image quality will be poor. On the other hand, if it waits too long after the clouds have disappeared then precious time during which there is no occlusion will be wasted aiming the instrument instead of taking images. The aiming procedure can be controlled by the mission manager and it can take anywhere between 2 and 5 units of time. An ideal duration is 3 or 4 units, since a short time of 2 units would put the instrument under pressure, while a long duration, like 5 units, would waste energy.

This scenario, rather tedious to describe in words, can be compactly represented by the STPPU shown in Figure 3.11 with the following features:

- a set of executable time-points SC (Start Clouds), SA (Start Aiming), EA (End Aiming);

- a contingent time-point EC (End Clouds);

- a set of soft requirement constraints on $\{SC \rightarrow SA, SA \rightarrow EC, SA \rightarrow EA\}$;

- a soft contingent constraint $\{SC \rightarrow EC\}$;

- the fuzzy semiring $S_{fuzzy} = \langle [0, 1], \max, \min, 0, 1 \rangle$.

As always, a solution of an STPPU P is a **schedule** T, that is, a complete assignment of values to all time points $T : (N_e \cup N_c) \rightarrow \mathbb{R}$, satisfying all the constraints. The set of all solutions is denoted by $Sol(P)$. A schedule consists of two parts: the **situation** (usually denoted by ω) is the tuple of durations identified by the schedule on the contingent constraints, and the **control sequence** (usually denoted by δ) is the tuple of assignments to all executable points. Given an STPPU P, a **projection** P_ω is the STPP obtained by assigning the time points in ω to P and $opt(P_\omega)$ is the preference of an optimal solution of P_ω; Proj(P) includes all P_ω. A **strategy** $S :$ $Proj(P) \rightarrow Sol(P)$ is a map from a projection to a schedule; if it is such that $\forall P_\omega S(P_\omega)$ is a schedule that includes ω, then it is said to be **viable**. The time assigned to executable variable x by schedule $S(P_\omega)$ is denoted by $[S(P_\omega)]_x$.

In [97] the notions of controllability are extended to accommodate preferences to provide an agent with the ability to execute the time-points under its control, not only subject to all constraints but also in the most preferred way. This entails a re-interpretation of the concept of optimality due to the presence of uncontrollable events. In fact, the distinction on the nature of the events induces a difference on the meaning of the preferences expressed on them, as mentioned in the previous section. Once a scenario is given it will have a certain level of desirability, expressing how much the agent likes such a situation. Then, the agent often has several choices for the events he controls that are consistent with that scenario. Some of these choices might be preferable to others. This is expressed by the preferences on the requirement constraints and such information should guide the agent in choosing the best possible actions to take. Thus, the concept of optimality is now "relative" to the specific scenario. The final preference of a complete assignment is an overall value which combines how much the corresponding scenario is desirable for the agent and how well the agent has reacted in that scenario.

The concepts of controllability presented in [97] are, thus, based on the possibility of the agent to execute the events under her control in the best possible way given the actual situation. Acting in an optimal way can be seen as not lowering further the preference given by the uncontrollable events.

The strongest notion of controllability, i.e., strong controllability, is extended in [97], in two ways, obtaining Optimal Strong Controllability (OSC) and α-Strong Controllability, where $\alpha \in A$ is a preference level. The first notion corresponds to a stronger requirement, since it assumes the existence of a fixed unique assignment for all the executable time-points that is optimal in every projection. The second notion requires such a fixed assignment to be optimal only in those projections that have a maximum preference value not greater than α, and to yield a preference $\not< \alpha$ in all other cases.

In other words, an STPPU is OSC if there is a fixed control sequence that works in all possible situations and is optimal in each of them. In the definition, "optimal" means that there is no other assignment the agent can choose for the executable time-points that could yield a higher preference in any situation. Since this is a powerful restriction, as mentioned before, one can instead look at just reaching a certain quality threshold, and aim at α-*Strong Controllability*.

An STPPU is α-Strongly Controllable if there is a fixed control sequence that works in all situations and results in optimal schedules for those situations where the optimal preference level of the projection is less than α and in a schedule with preference not smaller than α in all other cases.

In [97] an algorithm, called BestSC, is presented for checking whether an STPPU is OSC. If it is not OSC, the BestSC detects this and returns the highest preference level α such that the problem is α-SC. BestSC relies on the following tractability assumptions, inherited from STPPs: (1) the underlying semiring is the fuzzy semiring S_{fuzzy}, (2) the preference functions are semi-convex, and (3) the set of preferences [0, 1] is discretized in a finite number of elements according to a given granularity.

BestSC is based on a simple idea: for each preference level β, it finds all the control sequences that guarantee strong controllability for all projections such that their optimal preference is $\geq \beta$, and optimality for those with optimal preference β. Then, it keeps only those control sequences that do the same for all preference levels $> \beta$. Notice that, in contrast with finding an optimal solution of an STPP, it is not possible to use a binary search over preference levels (in contrast to algorithms for STPPs), since the correctness of the procedure is based on the intersection of the result obtained at a given preference level, γ, with those obtained at *all* preference levels $< \gamma$. Thus, determining the OSC or the highest preference level of α-SC of an STPPU with n variables and ℓ preference levels can be achieved in time $O(n^3\ell)$.

Similarly, the least restrictive notion of controllability has been extended to preferences by requiring the existence of an optimal solution in every situation. In other words, an STPPU is *Optimally Weakly Controllable* (OWC) if, for every situation, there is a control sequence that results in an optimal schedule for that situation. It easy to see that the Optimal Weak Controllability of an STPPU is equivalent to the Weak Controllability of the corresponding STPU obtained by ignoring preferences. The reason is that if a projection P_ω has at least one solution then it must have an optimal solution. This means that in order to check OWC of an STPPU it is sufficient to apply the algorithm described in Section 3.2.2 to the STPU obtained by simply ignoring all the preference functions. Moreover, any STPPU is such that its underlying STPU is either WC or not. Hence it does not make sense to define a notion of α-Weak Controllability.

As noted earlier, Dynamic Controllability (DC) addresses the ability of the agent to execute a schedule by choosing incrementally the values to be assigned to executable time-points, looking only at the past. When preferences are available, it is desirable that the agent acts not only in a way that is guaranteed to be consistent with any possible future outcome but also in a way that ensures the absence of regrets w.r.t. preferences.

Accordingly, an STPPU is *Optimally Dynamically Controllable* (ODC) if there exists a means of extending any current partial control sequence to a complete control sequence in the future in such a way that the resulting schedule will be optimal. As for the other controllability notions, the optimality requirement can be relaxed to having a preference reaching a certain threshold.

An STPPU is *α-Dynamically Controllable* if there is a means of extending any current partial control sequence to a complete sequence; but optimality is guaranteed only for situations with preference $\not\succ \alpha$. For all other projections the resulting dynamic schedule will have preference at not smaller than α.

The algorithm BestDC proposed in [97] to check ODC echoes BestDC for checking Optimal Strong Controllability. As done by BestDC, it considers the STPUs obtained by cutting the STPPU at various preference levels. For each preference level, first it tests whether the STPU obtained considering it as an STP is path consistent. Then, it checks if the path consistent STPU obtained is dynamically controllable, using the algorithm proposed in [76]. Thus, the control sequences that guarantee DC for scenarios having different optimal preferences are found. The next step is to select only those sequences that satisfy the DC requirement and are optimal at all preference levels.

The complexity of determining ODC or the highest preference level α of α-DC of an STPPU with n variables, a bounded number of preference levels ℓ is $O(n^5 \ell)$. Such a result, presented in [97], is unexpectedly good. In fact, it shows that the cost of adding a considerable expressive power through preferences to STPUs is a factor equal to the number of different preference levels. This implies that solving the optimization problem and, at the same time, the controllability problem, remains in P, if the number of different preference levels is bounded.

3.3.2 CONDITIONAL TEMPORAL PROBLEMS WITH PREFERENCES

The conditional nature of CTPs is enclosed in the variable labels, whose truth value enables or disables the presence of variables in the problem. Such labels indeed act as rules that select different execution paths, which, given variable v and its label $L(v)$, can be written as follows: IF $L(v)$ THEN EXECUTE (v).

The idea of fuzzifying such rules has been taken into consideration, for example in the field of fuzzy control [23, 65]. In fact, real world objects often do not present a crisp membership and classical Logics has difficulties to describe some concepts (e.g., "tall", "young", etc.). Another problem is that temporal information is often affected by imprecision or vagueness.

In a general study of such rules [31], both the premise and the consequence of the rule have been equipped with truth degrees associated with them. In [35] the same is done for the rules of a CTP. In such a case, however, these degrees have different meanings: the degree of the premise is used to establish if the variable should be executed, and therefore provides a truth value; the degree of the consequence, instead, can be considered as a preference on the execution of the variable. More formally,

Definition 3.9 CTPP. A CTPP is a tuple $\langle V, E, L, R, OV, O, \mathcal{P} \rangle$ where:

- $\mathcal{P}$ is a finite set of fuzzy atomic propositions with truth degrees in $[0, 1]$;

- V is a set of variables;

- E is a set of soft temporal constraints between pairs of variables $v_i \in V$;

- $L : V \to \mathcal{Q}^*$ is a function attaching conjunctions of fuzzy literals $\mathcal{Q} = \{p_i : p_i \in P\} \cup \{\neg p_i : p_i \in P\}$ to each variable $v_i \in V$;

- $R : V \to \mathcal{FR}$ is a function attaching a fuzzy rule $r(\alpha_i, cpn)$ to each variable $v_i \in V$;

- $OV \subseteq V$ is the set of observation variables;

- $O : \mathcal{P} \to OV$ is a surjective function that associates an observation variable to each fuzzy atomic proposition. Variable $O(A)$ provides the truth degree for A.

Each fuzzy rule $r(\alpha, cpn)$ states that variable v to which it is associated, is part of the problem if the minimum fuzzy interpretation associated to any fuzzy literal in its label is greater than the threshold α. Moreover, the consequence specifies the preference associated with the execution of v. In general, such a preference can depend on the truth degree of the premise and on the time at which v is executed. Therefore, it is reasonable to define $cpn : [0, 1] \to (R^+ \to [0, 1])$, that is, as a function which takes as input the truth degree of the premise, and returns a function which, in turn, takes as input an execution time and returns a preference in $[0, 1]$. In other words, function cpn allows us to give a preference function on the execution time of v which depends on the truth degree of the label of v.

The definitions of scenario, projection, schedule, and strategy are analogous to the classical counterparts. A *scenario* s chooses a value for each fuzzy literal, and, thus, determines which variables are executed and also which preference function must be used for their execution. This means that a scenario *projection* $Pr(s)$ contains the executed variables, the temporal constraints among them, and the information given by the preference function of each of the executed variables. A *schedule* $T : V \rightarrow R^+$ of a CTPP P is an assignment of execution times to the variables in V. Given a scenario s and a schedule T, the preference degree of T in s is $pref_s(T) = \min_{c_{ij} \in Pr(s)} f_{ij}(T(v_j) - T(v_i))$, where f_{ij} is the preference function of constraint c_{ij} defined over variables v_i and v_j.

Example 3.10   Figure 3.13 shows an example of CTPP, taken from [35], and representing a scenario where there is a rover exploring a certain region. The rover must leave its current position H and get to site W where a slope starts. Depending on how steep the slope is (this will become known at W), the rover can either go up the slope and reach the top T, or it can reach the top T by using additional power from an extra motor M, or it can turn around and go to another site of interest R to take some pictures. If the rover can climb the hill without the need of extra power, it should do so as fast as possible, reaching the top before 11 a.m. Such a constraint does not apply if the additional motor is needed, since, if the robot goes too fast, it might run out of energy before the top is reached. If it is not possible to get to the top of the hill, then the rover should get to site R only after 1 p.m. The later the better, since the light will be best for taking the pictures.

Such a scenario can be represented by a CTPP with nine variables: the start of time, x_0, the start and end time of the trip to W, denoted with HW_s and HW_e, the start and end time of the trip from W to T without extra power, denoted with WT_s and WT_e, the start and end time of the trip from W to T with extra power, denoted with WM_s and WM_e, and the start and end time of the trip from W to R, denoted with WR_s and WR_e. There are three fuzzy propositions, A, B and C, respectively representing the three fuzzy concepts "the slope is gentle", "the slope is steep", and "the slope is very steep". In Figure 3.12 we show the truth values corresponding to the three propositions (y-axis) in terms of the percent slope (x-axis). Variables WT_s and WT_e are labeled with proposition A and rule $r(0.4, cp)$, meaning that the rover should climb the slope without extra power only if the slope is gentle enough, that is only if the truth value of A is greater than 0.4. Similarly, for WM_s and WM_e, and WR_s and WR_e. At the same time, however, the more it is true that the slope is gentle (resp., steep, and very steep), the more preferable it is to get to the top (resp., use the motor, and go to site R). For this reason, the *cpn* functions of the rules are "directly" proportional to the truth degree of the observations, while making conservative choices more preferable in case of doubt. For example, we could have $cpn_1(x) = x$, $cpn_2(x) = \min(x + 0.1, 1)$ and $cpn_3(x) = \min(x + 0.2, 1)$.

The two temporal constraints of the example from x_0 to WT_e and to WR_e have been fuzzy-fied by using trapezoidal preference functions. The preference functions for the other constraints have been omitted, meaning that they are constant functions always returning 1.

Let us consider the case in which the percent slope is 10%. By looking at Figure 3.12 we see that this corresponds to a scenario, say s_1 such that $s_1(A) = 1$, $s_1(B) = 0$ and $s_1(C) = 0$.

Thus, projection $Pr(s_1)$ is the STPP defined on variables $x_0, HW_s, HW_e, WT_s, WT_e$. If instead the percent slope is 66%, we have a scenario, say s_2, such that $s_2(A) = 0$, $s_2(B) = 0.5$ and $s_2(C) = 0.5$, and projection $Pr(s_2)$ defined on variables $x_0, HW_s, HW_e, WM_s, WM_e, WR_s, WR_e$.

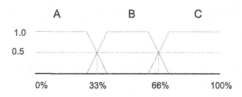

Figure 3.12: Truth values of the fuzzy propositions of Example 3.10.

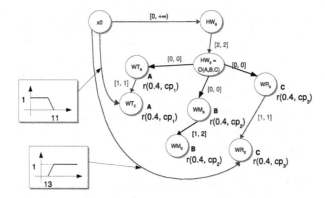

Figure 3.13: Example of Conditional Temporal Problem with Preferences.

Consistency notions in CTPPs are analogous to the ones in CTPs. However, preferences must be taken into consideration. For a CTPP to be α-*strongly consistent* (α-SCS), we must have a schedule that satisfies all the constraints independently of the observations, giving a global preference greater than or equal to α. This is the strongest consistency notion since it requires the existence of a single schedule that gives preference at least α in every scenario. On the contrary, we can just require the existence for every scenario of a schedule (possibly a different one for different scenarios) that has a preference of at least α given the corresponding projection. This notion is that of α-*weak consistency* (α-WCS). The above definitions are at the two extremes with respect to assumptions made on which events will be executed: α-SCS assumes no knowledge at all, while α-WCS assumes the scenario is given. A notion of consistency which lies in between is α-*dynamic consistency* (α-DCS) which has been considered in depth in [34]. It assumes that the information on which variables are executed becomes available during execution in an online fashion.

More precisely, a CTPP is α-DCS if for every variable v, whenever two scenarios (s_1 and s_2) are not distinguishable at execution time for v there is an assignment to v which can be extended to a complete assignment which in both scenarios will have preference at least α. It is easy to see that, as for CTPs, α-SCS $\Rightarrow \alpha$-DCS $\Rightarrow \alpha$-WCS. Moreover, given $\alpha \in [0, 1]$, if an CTPP is α-SCS/DCS/WCS then it is β-SCS/DCS/WCS $\forall \beta \leq \alpha$. Algorithms for testing α-SCS and α-WCS have been proposed in [35] whereas in [36] the focus is on enforcing α-DCS.

As mentioned above, for a CTPP to be α-strongly consistent, there must be a schedule that satisfies all the constraints independently of the observations, giving a global preference greater than or equal to α. In [35] a simplified case is considered first. In particular, a procedure for checking the α-SCS of CTPPs where the preference functions of the rules are independent of the truth degree of the label is given. The main idea behind such a procedure is to construct an appropriate STPP, associated with the input CTPP, such that its optimal consistency level corresponds to the highest level α at which the CTPP is α-SCS. The procedure is then extended to generic CTPPs where constraints may have different preference functions depending on the truth level of the label. In that case, the authors show that it is sufficient to consider the function which is defined by taking the minimum of all the functions on the constraints and then to apply the same above procedure.

As we have mentioned earlier, in classical CTPs, the problem of checking WCS is a co-$\mathcal{NP}$ complete problem [111]. Therefore, since the CTPP is an extension of the CTP, we cannot expect to do better. The algorithm proposed in [35] tests α-WCS for the restricted class of CTPPs defined above. In particular, it computes a minimal and exhaustive set of scenarios and then for each such scenario, say ms, it checks if the corresponding projection $Pr(ms)$ is consistent at level α. If the preference functions are semi-convex, in order to test this it is sufficient to test whether the STP obtained from $Pr(s)$ via its α-cut (that is considering for each constraint the sub-interval containing elements mapped into a preference $\geq \alpha$) is consistent which takes polynomial time. The number of scenarios checked may be, however, exponential.

Finally in [36] a procedure for checking the α-DCS of a CTPP is proposed. The algorithm is structured into three steps. The first step coincides with what is done in the case of strong consistency, that is a new CTPP in which preference functions are independent of the truth values of the labels is extracted by taking for each constraint the function which is the minimum of all the functions on that constraint. Then, as for the case for weak consistency, a minimal and exhaustive set of scenarios is computed. Finally, similarly to the case without preferences the problem is reduced to that of finding an optimal solution of a Disjunctive Temporal Problem with Preferences (DTPP). The complexity of testing the α-DCS of a CTPP remains an open problem, as has that of testing DCS for CTPs [36].

3.4   SUMMARY

In this chapter we have reviewed how constraint-based frameworks for reasoning about time can be extended to incorporate information on preferences and to model uncertain scenarios. We have first described how preferences can be added to qualitative models, as in the case of the fuzzy extensions of Allen's algebra, the point algebra, and their subclasses [6, 32, 84, 86]. In such extensions preferences are numbers in the [0,1] interval associated to primitive relation in the constraints. The higher the preference the more preferred is the relation, and, given a solution, the preference associated with it is the minimum among the preferences of the primitive relations it identifies.

Next we have considered the fuzzy extension of quantitative models [55, 57], where preferences are associated by preference functions to the elements of the intervals. In both cases we have seen that there are fundamentally two approaches to finding optimal solutions. The first one is to decompose the problem into several problems without preferences, while the other is to extend propagation rules to work directly with preferences. In either case, the idempotency of the aggregation operator, i.e., min, of fuzzy preferences and some other reasonable restrictions allow us to maintain the optimization problem in the same complexity class as the underlying hard constraint problem.

We have also mentioned how the rather coarse distinction that fuzzy preferences induce on solutions has been refined in the quantitative case by considering Pareto Optimal solutions [56] where solutions are ordered by looking at vectors of preferences. Moreover, quantitative frameworks have been extended with utilitarian preferences [75, 89], which model the more natural approach to preferences where the goal is to maximize the sum. Table 3.14 summarizes the main characteristics of constraint-based approaches to temporal reasoning with preferences.

We have then considered how uncertainty can be modeled within constraint-based formalisms. Uncertainty can occur in different forms. In STPUs [115], a quantitative approach, the uncertainty is on the occurrence time of some of the events. The notion of consistency is mapped into that of controllability, that is, the ability to schedule controllable events ensuring some degree of robustness with respect to the uncertain events. Different degrees of robustness correspond to different types of controllability, i.e., strong, weak, and dynamic, requiring the choice made for the controllable events to be consistent, respectively with all, one or future uncontrollable events. Different types of controllability have also different complexities. In fact, while strong and dynamic controllability are tractable [76, 79, 115], weak controllability is not [78].

While all the events corresponding to some variables in STPUs eventually occur, in CTPs the occurrence of an event is itself uncertain and modeled through the satisfaction of logical formulas [111]. The notion of consistency is extended to accomodate this: strong consistency requires a solution regardless of which events will actually occur, weak consistency requires the existence of a way of choosing the events so that there is a solution and dynamic consistency requires the solution to be built online in a backtrackfree fashion. This type of uncertainty can be overlaid on top of the different types of quantitative temporal problems, e.g., STPs and DTPs, and the complexity is basically inherited from the underlying structure [111].

Both STPUs and CTPs have been extended with preferences [34, 35, 36, 97]. In this type of problems solutions are evaluated both in terms of their robustness to uncertainty and quality with

	name	approach	temporal reference	pref type	temporal propositions	complexity
PA^{fuz}	fuzzy point algebra	qual	time points	(max,min)	$\{<[\alpha_1], = [\alpha_2], > [\alpha_3]\}$, $\alpha_i \in [0,1]$	tractable
PA_c^{fuz}	fuzzy convex point algebra	qual	time points	(max,min)	as PA^{fuz} with $\alpha_2 \geq \{\alpha_1, \alpha_3\}$	tractable
IA^{fuz}	fuzzy interval algebra	qual	intervals	(max,min)	$\{b [\alpha_1], m [\alpha_2], o [\alpha_3], s [\alpha_4], d [\alpha_5], f [\alpha_6], e [\alpha_7], bi [\alpha_8], mi [\alpha_9], oi [\alpha_{10}], si [\alpha_{11}], di [\alpha_{12}], fi [\alpha_{13}]\}$	NP-hard
IA_c^{fuz}	fuzzy convex interval algebra	qual	intervals	(max,min)	conjuctions of PA_c^{fuz} constr.	tractable
IA_p^{fuz}	fuzzy pointisible IA	qual	intervals	(max,min)	conjuctions of PA^{fuz} constr.	tractable
STPP	simple temporal problem with preferences	quant	time points	(max,min)	binary difference, semi-convex pref. functions	tractable
STPP-util	simple temporal problem with utlititarian prefs.	quant	time points	(max,+)	binary difference, piece-wise linear pref. functions	tractable
DTPP	disjunctive temporal problem with preferences	quant	time points	(max,min)	n-ary disjunctive difference, semi-convex pref. functions	NP-hard

Figure 3.14: An overview of main temporal frameworks with preferences.

respect to preferences. Table 3.15 provides a summary on constraint-based temporal formalisms which model uncertainty.

	name	uncertainty	pref type	Controllability/Consistency		
				strong	dynamic	weak
STPU	simple temporal problem with uncertainty	occurrence time	no preferences	tractable	tractable	Co-NP-c
CTP	conditional temporal problem	occurrence	no preference	tractable	NP-c	Co-NP-c
STPPU	simple temporal problem with preferences and uncertainty	occurrence time	(max,min)	tractable	tractable	Co-NP-c
CTPP	conditional temporal problem with preferences	occurrence	(max,min)	tractable	NP-c	Co-NP-c

Figure 3.15: An overview of main temporal frameworks with uncertainty. All approaches are quantitative. Complexity for CTPs and CTPPs are given assuming an underlying STP.

CHAPTER 4

Applications of Temporal Reasoning

4.1 INTRODUCTION

In Chapter 1 we introduced a model for an *agent-in-time*, one that continuously senses, decides, and acts in a world that changes over time. The subsequent chapters introduced different representations of time for AI and agent-based systems, as well as algorithms for solving reasoning problems involving these representations. This discussion has laid the foundation for the current chapter, which will illustrate how these temporal systems of representation and reasoning can be embedded in larger systems and applied to solve real world problems.

We subdivide this chapter as follows. First, we distinguish broadly between temporal planning applications and applications involving the temporal interpretation of data. Temporal planning (including scheduling and plan execution) is the process of taking a set of goals, an action model and a set of initial world conditions and generating an assignment of actions in time that satisfy the goals. By contrast, the temporal interpretation of data starts with (structured or unstructured) time-stamped data, and extracts temporal information from it. Within data interpretation applications we distinguish between real-time data interpretation and interpreting from stored information, such as databases or texts. In the planning sections we focus on applications for space exploration, but in the data interpretation section we include applications from many domains.

4.2 ACTIVITY PLANNING

In the classical statement of the problem, a planning system takes as inputs a set of activities or *goals* (roughly, conditions of the world that need to be achieved), and a description of the initial state of the world. The output is a plan, an ordered collection of actions that collectively achieve the goals. The planning task also involves a set of rules or constraints that define the tasks and their interactions with the world. Examples of these rules include rules for decomposing goals into tasks, or decomposing tasks into sub-tasks; rules that define pre-conditions that must be true for an task to be performed, and post-conditions that describe how a task alters the world; and rules that define the interactions between tasks and resources that use them (e.g., between driving and the consumption of fuel).

4.2.1 CONSTRAINT-BASED PLANNING

Constraint-based Planning [41] defines a family of planning systems that consists of the following elements:

- Time as a network of variables and relations, typically, a variant of a network associated with an STP (as defined in Section 2.2.1).

- A set of *timelines* describing the history of a world state variable of interest over time.

- A network of states and activities describing the evolution of the plan over time; and

- A planning algorithm based on the incremental resolution of *flaws*, for example, unplanned activities or constraint violations.

A full detailed introduction to automated activity planning in general, or constraint-based planning in particular, is beyond the scope of this book. Our focus instead will be how frameworks of time described earlier in this book are embedded into constraint-based planning systems.

Examples of constraint-based planners that have been used to plan for science activities in space are EUROPA [14], ASPEN [18], and IxTeT [12]. Examples of space applications of these planners include EO-1 on-board science planning [104]; MEXAR download operations planning [22]; Deep Space Network Scheduling [51]; MER Science Mission Planning [1], and Deep Space 1 planning [83].

Planning for space activities is difficult because the requisite plans involve large numbers of concurrent tasks with temporal dependencies, limited resources, and complex rules defining interactions among physical systems. Often there are critical computational limitations due to deadlines, and temporal reasoning must be fast and flexible. Finally, often activity planning for space science is "over-constrained:" there are more goals than can be achieved given the constraints, and a solution must choose a subset of goals to be achieved based on a notion of priority.

In planning, the underlying temporal CSP is usually *dynamic*: variables or constraints are added or deleted over time as the plan evolves. We've seen that temporal constraint reasoning can be used to find solutions (consistent assignment of values to variables); to determine the consistency of a problem, and to eliminate possible values to variables (constraint propagation). In planning these operations are integrated into a plan search algorithm, which incrementally develops a plan either forward from the initial state, or backward from the goal state, to build a complete feasible plan.

4.2.2 EXAMPLE PLAN

For an illustration of the main components of a plan produced by a constraint-based interval planner, and issues that emerge related to time, Figure 4.1 shows a fragment of a plan for an instrument on a spacecraft to take a picture of an asteroid (AST1). The primary tasks involve changing spacecraft attitude so that it is pointing at the asteroid, and turning the camera on so

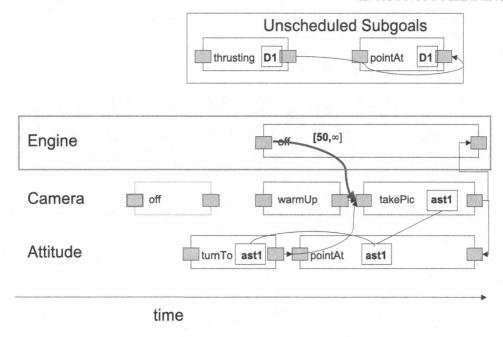

Figure 4.1: Example of a partially instantiated constraint-based interval plan.

that it can take a picture. Although simple, the plan illustrates all the features of a constraint-based plan, especially the temporal features. First, an activity is defined over an interval that can either be a condition that is true (*the camera is off*) or a task (*warm up* the camera). The yellow box marked *off* in the figure is an example of an interval with start and end points (the boxes in red). The figure associates the interval to the *camera* state variable, meaning it is a value of that variable for the designated duration. Furthermore, some activities, such as the *takePic*, have arguments that take on values; in this case, the activity has the value AST1 indicating that the camera is taking a picture of AST1.

Second, a sequence of activities associated with the same state variable is called a *timeline*; thus the green box over the state variable "engine" is a timeline; it shows the evolution or history of state variable over time (in a final plan, all times on the timeline must be associated with some state for each variable). Third, the directed edges (arrows) depict temporal constraints between the endpoints of the activities. The edges marked with ranges $[a, b]$ are quantitative temporal constraints (see Section 2.2.1); the unmarked edges are qualitative. For example, the blue arrow specifies the constraint that the start of the engine must be off at least 50 time units before the start of the camera taking a picture. The plan also shows non-temporal constraints on the values of arguments of activities; for example, that the spacecraft must be pointing to the same object that the camera is taking a picture of. Finally, the plan shows unscheduled goals (i.e., a flaw) in the plan, consisting of a thrusting action of the engine.

4.2.3 FIXED VS. FLEXIBLE PLANS

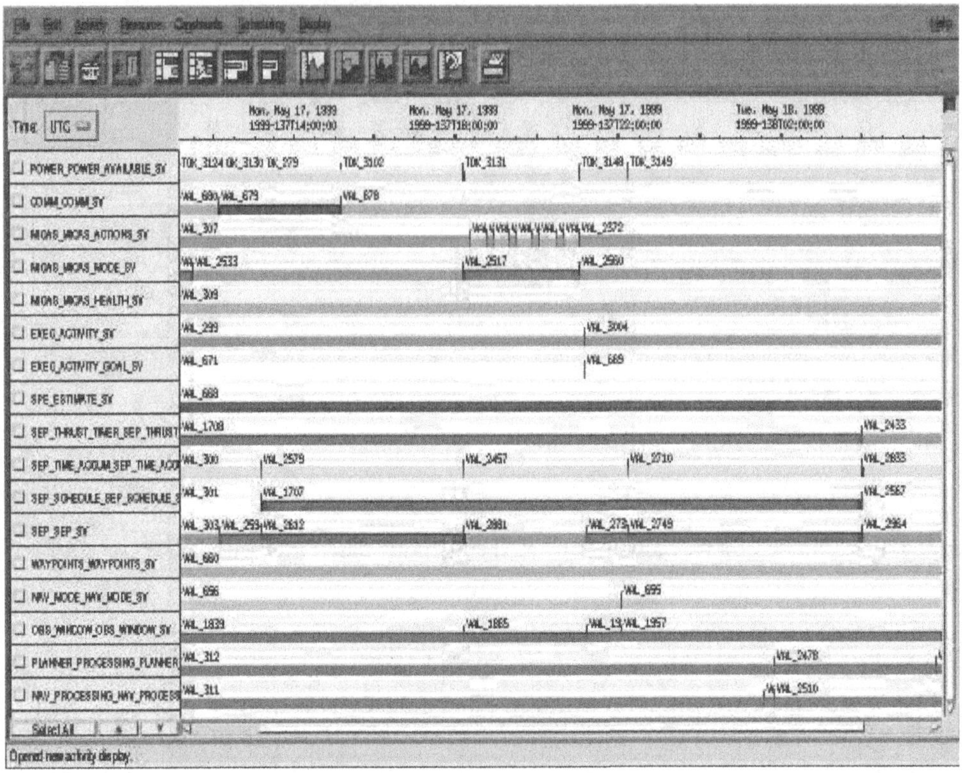

Figure 4.2: MAPGEN interface.

The underlying temporal structure of a completed constraint-based plan is a simple temporal network, with nodes corresponding to the end points of intervals in the plan. As we've noted, flexibility is important in dealing with temporal uncertainty. The end user, however, i.e., the mission activity planner, always prefers to be presented with a fixed plan, a set of timelines in which all the end points of the activities are assigned single values. Figure 4.2 shows the user interface for the MAPGEN activity planner for the MER mission, consisting of timelines and activities with fixed start and end times. One issue that arises in practice is *which* fixed plan should the planning system generate? One option is to present the plan where all activities are assigned their earliest start times. This is provably a valid plan, but from the user perspective it is not always the preferred plan. A related issue is how to change the fixed plan as more tasks are added. Both these issues involve not just maintaining the consistency of a plan but also what assignments and updates are useful and intuitive to the human user.

A practical solution is to allow users to express temporal preferences for start times in the fixed plan through the interface. Then, if new activities are added to the plan, the system tries to enforce user preferences on times of the fixed plan after propagation, using a notion of *minimum plan perturbation*. For example, suppose a new activity has been added, and the result of propagation has changed the start times of another activity to $[lb, ub]$. Further suppose that the preferred time specified by the user prior to the update is t, and that $t < lb$, so that the preferred time is no longer consistent. Then to enforce minimum perturbation of the plan, the preferred time is reset to lb, the closest to the user preferred time.

Constraint-based planning has been used successfully in many space applications. The benefits of using automated planning are hard to quantify, but MER mission operators have estimated that the amount of science that was achieved through the use of the MAPGEN has increased by 15 to 40 per cent, while reducing the planning time to a small fraction of the nominal time.

4.3 AUTONOMOUS EXECUTION FOR SPACE EXPLORATION SYSTEMS

Space exploration has triggered significant research toward the creation of exploration systems (primarily spacecraft and ground robotic explorers) with increased levels of *autonomy*. Autonomous vehicle behavior includes the ability for the vehicle to carry out high level mission goals with little or no human intervention while maintaining its safety in the presence of an unknown and often hostile environment [52].

Virtually all architectures for autonomy distinguish between a *deliberative layer* and an *execution layer*. The former can be viewed roughly as a continuous planning system in the sense described above, where continuous means that the planner is always generating or updating plans. The execution layer provides all the capabilities required to control the vehicle's sensors, effectors, and communication devices. The functionality required for an interface between the deliberative layer and the local control devices is called the *executive*: its function is to coordinate and control the operations at the subsystem level (sometimes called the *functional layer*) according to the requirements of the individual task [27]. Many designs for the execution layer include temporal constraint reasoning, and in this section we briefly examine this integration with two example execution systems.

4.3.1 REMOTE AGENT EXECUTIVE

Remote Agent (RAX) [82] on the Deep Space One experimental mission was the first artificial intelligence system to control a spacecraft without human supervision. Remote Agent successfully demonstrated the ability to plan onboard activities and correctly diagnose and respond to simulated faults in spacecraft components.

The RAX Executive [81] (EXEC) plays the main mechanism for coordinating the other primary software modules of the Remote Agent. The two main aspects of its behavior are comprised of periodically getting a new plan from the planning module, and executing the plan, guided by the state information provided by the model-based state estimation component.

The EXEC's algorithm interprets a constraint-based interval plan by assigning the actual run time of the activities on the plan's timelines, and propagating execution time information through the plan's temporal network. More specifically, each state variable in the plan is treated as a thread of execution, and each activity on the timeline is a program to be run at the start time indicated on the timeline. A list of timepoints is maintained for activities that become available for execution as time passes and as other dependent activities are executed. When a time point is enabled and the activity selected for execution, the following steps are performed [81]:

1. the execution point is set to the current time;

2. the activities started by the selected activity are also started, thus resetting their start times;

3. the activities ending with the start of the selected activity are ended, thus resetting their end times;

4. the changes in time are propagated.

One issue involving time that arises is that of *latency*: the potential delay in execution as a result of the running the plan runner loop. For example, if it takes (in the worst case) λ time units to complete a cycle of steps 1–4 above, and an activity scheduled for time t_0 becomes enabled, then there's the potential for delay in execution as some fraction of $t_0 + \lambda$. The distinction between an activity's scheduled start time and the actual time it is executed is important in both the design and execution of temporal plans. For plan design, it is important not to schedule two activities with a separation constraint less than λ. One way of reducing λ is to speed up constraint propagation, which has been accomplished through transforming the STN into a minimal dispatchable network, as discussed in Section 2.2.2.

Lessons learned from NASA's RAX have led to new concepts, languages, and architectures for autonomy that incorporate a more robust model of time for execution. The dual conflicting requirements of reactivity (timely successful execution of all actions) and robustness (successful response to unexpected events) have led to multiple approaches to designing autonomous systems based on constraint reasoning that satisfy these requirements. One such solution is discussed in the next subsection.

Along the spectrum of design choices the RAX executive can be viewed as emphasizing reactivity; "robustness" in EXEC consisted of "putting the spacecraft in safe mode and asking the planning system for a new plan" [81]. More recent designs, including the following example, enforce more sophisticated forms of robustness through the migration of decision-making, including decisions involving time, from the planning phase to the execution phase. The RAX Executive can also be viewed as "operator"-centered, whereby a science or engineering team builds a temporal plan by adding activities, using an interface such as the one described above. Once the plan is verified, it is compiled into an execution sequence that can be uploaded to the spacecraft for autonomous execution.

4.3.2   REACTIVE MODEL-BASED PROGRAMMING LANGUAGE (RMPL)

The Reactive Model-based Programming Language (RMPL) is a framework for representing complex system behaviors. Unlike the RAX EXEC, the overall approach is "programmer-centered:" the starting point of its design are more traditional embedded programming languages for controlling complex systems that allow for the programming of low level behaviors such as operating a sensor, as well as enforcing concurrency and preemptive behaviors. To this, RMPL adds capabilities for reasoning, including temporal reasoning, that are similar, but to some extent extend, the capabilities of the RAX executive [58]; hence RMPL is a "unification" of reactive with model-based (constraint-based) programming.

Like the RAX EXEC, RMPL has a representation of time in order to enforce metric timing requirements and temporal dependencies between activities and conditions, along the lines of the constraint-based planning approach described earlier. Specifically, an RMPL program is compiled into a *Temporal Plan Network* (TPN) for execution. A TPN is a variation of the TNA introduced in Section 2.2.4: an STN augmented with *symbolic constraints* and *decision nodes*. Roughly, symbolic constraints allow for the expression of conditions that must be true for an activity to be executed, and decision nodes represent choice (branch) points to allow redundant activities. To this extent, the plan representation allows for conditional execution in a way that is not present in the timelines model. An alternative interpretation of an RMPL program is as a Hierarchical Timed Automata (HTA), which provides a more explicit representation of the semantics in terms of states and timed transitions [33].

The primary argument for more temporal planning during execution is robustness. The reason for this is that information is being accumulated during execution, which reduces the uncertainty for the agent, hence enabling more robust behavior. In a programmatic-centered approach, execution failure can be defined as *uncaught exceptions* during executions (i.e., exceptions which the code is not equipped to handle). From a temporal planning perspective, failures can be timing constraint violations (e.g., something not started in time) or activity failures (an activity not terminating normally).

A recent set of extensions of the RMPL framework to enhance robustness is based on three RMPL compilation techniques and two activity models, which when combined offer a suite of execution strategies that trade exactness with speed and scalability [33]. The problem is formulated as that of minimizing risk (of uncaught exceptions), but it is related to the notions of temporal plan controllability (especially dynamic controllability; see Section 3.2.1).

One strategy is based on probabilistic models of activities, representing both nominal and activity failure as probability distributions over time. For example, a stochastic model of the duration of an activity might be defined as a Gaussian distribution on a bound $[a, b]$. Similarly, when the same activity fails, the times at which it fails may define a separate distribution, with a higher probability (say) at the onset of the activity, and roughly uniform thereafter. A Markov Decision Process (MDP) can then be generated from the HTA and set of stochastic activity models, which can be solved using value iteration, resulting in a minimal risk decision policy.

An alternative approach, based on dynamic controllability, especially for Conditional Temporal Problems [112], is called set-bounded dynamic controllability. The foundation of this approach is a set bounded activity model, which represents temporal bounds as either controllable or uncontrollable, as defined in Section 3.2.1. From an executive's point of view, a controllable activity duration is something scheduled, whereas an uncontrollable activity duration is something that can only be observed during run time. A TPN is extended to Temporal Plan Network with Uncertainty (TPNU). Dynamic controllability of a TPNU is represented as a two-person game between an agent and the environment, where the role of the environment is to decide when an uncontrollable event begins or ends, and also which program exceptions to throw. The set of possible games is represented as an AND-OR search tree. Dynamic controllability is the condition of the tree in which all possible paths satisfy the simple temporal constraints. The AND-OR tree is a representation of the execution strategy to ensure successful execution. See [33] for more details.

Executing temporal plans continues to be an active area of research in the design of automated robotic systems, from search and rescue applications [16], to undersea exploration for science [72]. The twin demands of reactivity and robustness continue to drive the design requirements of temporal reasoning during execution, and different domains suggest different solutions. Finally, there is interest in expanding temporal planning for execution to multiple agents [46], where issues arise relating to the effective partitioning of temporal networks into subnetworks, each executed by a separate agent.

4.4 EXTRACTING TEMPORAL INFORMATION FROM DATA

Our interest in this section is in methods for the extraction and inference of *temporal* patterns from sequences of data. Broadly, we can distinguish between real-time applications of temporal abstraction (where the data are streaming in real time) and applications in which timestamped data are stored in a database. We give examples of each below. With real time data the primary technical challenge for interpretation systems is *timeliness*: how to respond with information in time to be useful. With stored data the technical challenge is more often *data volume*: how to identify and summarize useful information from large quantities of data. In the remainder of this section, we illustrate what we will call temporal pattern abstraction in the following applications:

- Real-time situation recognition/monitoring;

- Question-answering systems;

- Temporal reasoning from natural language text; and

• 3d reconstructions of image data: inferring dates of an historical image;

4.4.1   REAL TIME SITUATION RECOGNITION/MONITORING

We first examine the use of time in embedded (also called situated) agents responding continuously to changes in the world. The first example is temporal information to support situational awareness of time for highly reactive environments such as fighter jet piloting. Here the amount of temporal reasoning and the information presented is highly constrained by the response-time requirements, and by the need to not overload the pilot with information. In [108], a visual timeline-based approach to temporal information is presented. The intervals on the time line represent the actions taken by adverse objects, and possible pilot responses. The intervals move along the screen toward the "now" and change color to indicate urgency or danger. The goal is to enhance or augment pilot awareness of the response options he has available now or in the near future. The data used as input include estimates from sensors of adverse object position, speed, and direction, and also predictive information about possible changes to those state values. It may also include classification of type of craft, weaponry, etc. From this the pilot, along with a decision support tool, can formulate offensive or avoidance actions.

A second example of continuous temporal reasoning by embedded systems is a bidding agent for buying and selling stock [28]. A client may say: *buy a stock S if there is a greater than 80% probability (based on output from a prediction program) that it will go up in the next 2 weeks.* Thus, temporal decisions include deciding the "right time" to perform a transaction to ensure a high probability of positive outcome to a client. In [28] a temporal reasoning framework is combined with reasoning under uncertainty to define a Temporal-Probabilistic (TP-) Agent. Somewhat like a plan execution system, a TP-agent has a set of actions it can execute, defined in the standard way in terms of preconditions and effects. The actions can be interpreted as calls to some external code (for example, $buy(S)$ is an action in this domain that calls external code to add S to a client's portfolio). Actions can be combined to form concurrent actions, and are annotated with temporal and probabilistic information that provide conditions under which the action is executed. In addition to actions, there are *status conditions* that describe states of the world, including the state of the client (e.g., how much stock he or she currently owns). A TP Agent Program is a set of rules that relate actions to status conditions: for example, a rule might say *If stock X is in my current portfolio and it is expected with 80% probability to go over $50 per share sometime in the next 10 to 20 days, then buy this stock in the next 5 days.*

A third application involving processing of continuous data is *chronicle recognition* to support surveillance by Unmanned Aerial Vehicles (UAVs) [44], [29], as well as potentially other monitoring tasks [30] . A chronicle is a complex scenario that consists of a set of temporally constrained events. Chronicle recognition helps to summarize potentially large amounts of temporal and spatial data: for example, a chronicle such as "passing a car along a highway" can allow a UAV to summarize large amounts of position and speed data as "Car A is passing car B". A chronicle can be defined as a set of events and their temporal relationships, i.e., a simple temporal network.

The chronicle recognition problem is that of inferring a chronicle (instance) from a collection of time-stamped data. A pre-processing step involves the construction of "primitive events" that are expressed in the language of the chronicle; for example, from raw position data a primitive event such as "Car A is next to car B (at t)" can be constructed, which is an event in the car-passing chronicle. The basic chronicle recognition algorithm consists of trying to *integrate* each incoming event into partial instances of the chronicle model, or creating a new chronicle instance and propagating the temporal bounds to other events in the model. For example, if an event (e, t) is observed and there is a chronicle constraint $f - e \in [2, 4]$ then a new chronicle instance can be created with temporal bounds set on an expected future occurrence of f. On this approach, partial chronicle instances are destroyed if an event predicted by the partial chronicle does not occur. More sophisticated approaches deal with the potential explosion of partial chronicle instances by a process known as focusing, which creates chronicles only if "uncommon" events happen (e.g., if f is much less common than e then the focus will be on waiting for f to occur before creating the chronicle [30]).

The final example of real-time temporal processing of data is plan execution monitoring. The set up of this capability is one agent tracking the progress of another agent executing a plan, reporting (to the execution agent or a third party) any trends that could lead to plan failure. To accomplish this the system must constantly update the plan being executed, and to develop a belief model about the agent executing the plan, in order to make inferences about possible future actions. We briefly describe a temporal constraint-based execution monitoring system in the domain of robotic assistance for the elderly [20]. As with the other examples in this section, we need a model of time that deals with a changing world state and uncertain information. The planning framework for this application is Disjunctive Temporal Networks (DTNs), as introduced in Section 2.2.3. The monitoring framework, called Quantitative Temporal Bayesian Networks (QTBN) is comprised of two components: a version of Time Nets (TNs) [25] for representing temporal uncertainty and temporal relations, and Dynamic Bayes Nets (DBNs) for representing non-temporal uncertainty and causal relations. Each component complements the other: TNs represent general temporal relations but have no simple mechanism for accommodating change; DBNs have a mechanism (rollup) for accumulating evidence but have limited expressive power for temporal constraints because of the Markov assumption. The interface between the two components allows for sharing temporal information. For example, the DBN needs temporal information about the current time slice from the TN. Conversely, the TN needs to update temporal beliefs from evidence stored in the DBN. A working version of the execution monitoring system has been integrated into the Autominder system [91].

4.4.2   3D RECONSTRUCTIONS OF IMAGE DATA

Given a set of old photos of, for example, an urban skyline taken from different angles (most, say taken before automatic digital dating), what is their ordering in time? Given an object that occurs in more than one photo, say a building, what can you infer about the construction/demolition date of that object? The problem posed by these questions is an extension of the standard Structure from Motion (SfM) problem, concerned with recovering the 3D geometry of the objects in a scene, as well as the geometry camera capturing the scene.

Formally, given a set of n images, registered to a set of m objects, we wish to assign values to a set of *temporal parameters* T, of two kinds: timestamps $\{t_i : 1 \leq i \leq n\}$ to the images, and durations: $\{[a_j, b_j] : 1 \leq j \leq m\}$ to the objects. In [101], the SfM problem is extended to derive these values using Bayesian reasoning. The inference is based on an observation model Z comprised of a set of associations of objects with images, and a geometric model X describing the geometry of the objects and the camera geometry in each image. Then the problem is to perform inference on all temporal parameters given observations and geometry:

$$P(T \mid Z, X) \propto P(Z \mid T, X) P(T) \tag{4.1}$$

The observation model from which the likelihood $P(Z \mid T, X)$ is derived is based on three factors: *viewability*, based on object and camera geometry; *existence*, based on evidence that the object existed at the time of the image; and *occlusion*, estimating whether the object is in view, which is a function based on geometry and time. The priors $P(T)$ are based on image date priors (which can be vague, such as that it is "circa 1960"), and object duration prior information.

The goal is to find an optimal assignment $T*$ for the temporal parameters: $T* = \mathrm{argmax}_T P(T \mid Z, X)$. In [101], a two step approach is described: first, the standard 3D SfM problem is solved using a traditional method; next a Markov Chain Monte Carlo algorithm is applied to solve the temporal reconstruction problem by sampling from the posterior distribution. Details and results are found in [101].

4.4.3   DETECTING TEMPORAL PATTERNS IN LARGE MEDICAL DATA SETS

The identification of temporal patterns from large data sets can aid in developing causal models, update plans, generate reports, and monitor and analyze complex processes. For example, temporal pattern identification in large medical patient databases is useful to monitor a patient treatment process given a known diagnosis [43]. The reasoning task starts from time-stamped patient state information (e.g., from tests, typically represented as parameter/value pairs such as GLUCOSE = high), knowledge about ongoing events (e.g., chemotherapy treatment), and an abstraction goal (e.g., monitoring an AIDS patient), and infers *temporal abstractions*, intervals that provide useful summaries to physicians who are seeking to monitor the patient's treatment and progress. An abstraction is defined by a type (state, gradient, or rate) and a time interval: an example is *Grade II anemia for 3 weeks in the context of administration of the drug AZT as part of clinical therapy protocol CCTG-522*. The (interpretation) context "administration of the drug AZT as part of clinical therapy protocol CCTG-522" is required for abstraction because such information can change the meaning or the values assigned to the abstraction.

In [17], a Constraint-based Pattern Specification Language (CAPSUL) is introduced to formulate temporal abstraction queries. Three kinds of constraints are defined: local, pertaining to single intervals (e.g., *a severe episode of anemia of 3 to 5 days*); global, that pertain to a relationship between intervals (e.g., *2 consecutive weeks of anemia where the second week was more severe than the first*; and repeating-pattern constraints that define temporal/value relations between repeating events: (e.g., *monthly episodes of anemia which contain at least 2 or more episodes in which the anemia was severe*). Local intervals are created by a number of operations on temporal objects (points or intervals) such as horizontal inference (joining together temporally overlapping intervals into a single interval); and temporal interpolation (joining points or intervals with gaps into a single interval). Global constraints are based on quantitative and qualitative temporal constraint languages discussed earlier in this book. Finally, repeating constraints requires additional operations on the data; for example the cardinality of the repetition (two or more episodes in which some value is observed), the overall duration of the pattern (look for a recurrence over a period of five months), and the desired value might be defined over non-convex intervals (i.e., ones with gaps); for example, the desired abstraction might be based on observing a decreasing time gap in a sequence of intervals of the same kind. A general calculus of non-convex intervals and constraints on repeating events is introduced in [80] and [66].

4.4.4   TEMPORAL INFORMATION EXTRACTION FROM NATURAL LANGUAGE TEXT

Interpreting temporal features in natural language has been an active research area for many years. Since the advent of the World Wide Web, a new resurgence of interest in the problem of *information extraction* has been driven by the need to distill useful knowledge from the vast body of natural language text found on the web. The temporal information extraction (TIE) problem is that of inferring temporal elements (points or intervals) and temporal constraints (preferably the tightest constraints inferable from the text) associated with events described in the text.

A number of different approaches to solving the TIE problem have been proposed. Most if not all use the qualitative or quantitative constraint-based representations of time discussed earlier; for example, TimeML [95] is a specification language for temporal markup based on Allen relations. Furthermore, since identifying temporal constraints can be viewed as a classification problem (e.g., identifying the point relation between times t_1 and t_2 associated with a pair of events in the text), many researchers have used supervised learning to build a classifier [109]. Others such as [67] use a form of probabilistic inference to extract point-wise constraints on the endpoints of event-intervals.

A simple but illustrative example of extracting information from text is temporal annotation. The goal of this task is to identify the flow of a piece of text by locating shifts in temporal focus in some extended narrative. For example, consider a fragment of a medical case summary: *A 12-year-old boy was evaluated in the hematology clinic of this hospital. For one week, he had had fatigue that had caused him to miss school, and his appetite had decreased. He also had intermittent cramping.* The first sentence introduces an event that happened in the past (an evaluation). The second sentence introduces a temporal jump to some time prior to that event. The third sentence then describes further events in the period introduced by the second (although there is no obvious temporal relation, such as overlap, implied for the events described in sentences two and three). The *segmentation classification task* is to decide where the temporal transitions occur in text. One feature in this example that indicates a transition between sentence one and two is the change in verb tense from simple past to past perfect "had had". A classifier can be applied to simultaneously find boundaries and imposes a temporal ordering in the boundaries using a Directed Acyclic Graph, from which other orderings can be then inferred by transitivity. In [13], temporal boundaries are feature vectors representing different textual markers of transitions. The classifier can score each boundary based on vector values, and then simultaneously identify the segments and orderings between them by casting the problem as an Integer Linear Programming (ILP) problem.

4.5   SUMMARY

The purpose of this chapter has been to illustrate the ubiquity of time in the design of artificial agents. In Chapter 1 we defined an agent as one that continuously senses, decides, and acts in a changing world. To act rationally, the agent must simultaneously maintain a world state model and agent plan from which it acts (see Figure 1.2). This chapter has presented a wide range of applications to which constraint-based representations of time can be applied in generating and executing plans, and in abstracting temporal patterns from data.

From designing artificial agents that can response to a continuously changing world, to building systems for extracting useful concise summaries from large data sets, graphical representations of time and temporal constraints provide an effective general framework for developing solutions to these challenging real world problems.

Bibliography

[1] M. Ai-Chang, J. L. Bresina, L. Charest, A. Chase, J. Cheng-jung Hsu, A. K. Jóns-son, B. Kanefsky, P. H. Morris, K. Rajan, J. Yglesias, B. G. Chafin, W. C. Dias, and P. F. Maldague. MAPGEN: Mixed-Initiative Planning and Scheduling for the Mars Exploration Rover Mission. *IEEE Intelligent Systems*, 19(1):8–12, 2004. DOI: 10.1109/MIS.2004.1265878. 82

[2] J. F. Allen. Maintaining knowledge about temporal intervals. *Communications of the ACM*, 26(1):832–843, 1983. DOI: 10.1145/182.358434. 17, 25, 41

[3] J. F. Allen. Time and time again : The many ways to represent time. *International Journal of Intelligent Systems*, 6(4):341–355, 1991. DOI: 10.1002/int.4550060403. 5

[4] A. Armando, C. Castellini, E. Giunchiglia. SAT-based procedures for temporal reasoning. In S. Biundo and M. Fox, editors, *Recent Advances in AI Planning (Proceedings of the 5th European Conference on Planning, ECP-99)*, volume 1809 of *Lecture Notes in Computer Science*, pages 97–108. Springer, 2000. 37, 40

[5] S. Badaloni and M. Giacomin. Flexible temporal constraints. In *8th Conference on Information Processing and Management of Uncertainty in knowledge-Based System (IPMU 2000)*, pages 1262–1269, 2000. 47

[6] S. Badaloni and M. Giacomin. The algebra IA^{fuz}: a framework for qualitative fuzzy temporal reasoning. *Artificial Intelligence*, 170(10):872–908, 2006. DOI: 10.1016/j.artint.2006.04.001. 46, 47, 77

[7] J. Bae, H. Bae, S.-H. Kang, Z. Kim. Automatic control of workflow processes using ECA rules. *IEEE Transactions on Knowledge and Data Engineering*, 16(8):1010–1023, 2004. DOI: 10.1109/TKDE.2004.20. 39, 40

[8] R. Barták and O. Čepek. Temporal networks with alternatives: Complexity and model. In *Proceedings of the Twentieth International Florida AI Research Society Conference (FLAIRS 2007)*, pages 641–646. AAAI Press, 2007. 37, 38, 40, 42, 67

[9] R. Barták and O. Čepek. Nested precedence networks with alternatives: Recognition, tractability, and models. In P. Traverso D. Dochev, M. Pistore, editors, *Artificial Intelligence: Methodology, Systems, and Applications (AIMSA 2008)*, volume 5253 of *Lecture Notes*

in Artificial Intelligence, pages 235–246. Springer Verlag, 2008. DOI: 10.1007/978-3-540-85776-1. 39, 40

[10] C. Bettini, X. S. Wang, and S. Jajodia. A general framework for time granularity and its applications to temporal reasoning. *Annals of Mathematics and Artificial Intelligence*, 22(1-2):29–58, 1998. DOI: 10.1023/A:1018938007511. 5

[11] S. Bistarelli, U. Montanari, and F. Rossi. Semiring-based constraint solving and optimization. *Journal of the ACM*, 44(2):201–236, mar 1997. DOI: 10.1145/256303.256306. 48, 52

[12] B. Bonet and H. Geffner. Planning as heuristic search. *Artificial Intelligence*, 1-2:5–33. DOI: 10.1016/S0004-3702(01)00108-4. 82

[13] P.J. Bramsen. *Doing Time: Inducing Temporal Graphs. Master Thesis*, Massachusetts Institute of Technology, Department of Electrical Engineering and Computer Science, 2006. 93

[14] J. L. Bresina, A. K. Jónsson, P. H. Morris, and K. Rajan. Activity planning for the mars exploration rovers. In S. Biundo, K. L. Myers, and K. Rajan, editors, *Proceedings of the International Conference on Automated Planning and Scheduling (ICAPS)*, pages 40–49. AAAI, 2005. 82

[15] M. H. Burstein and D. V. McDermott. Issues in the development of human-computer mixed-initiative planning systems. In *Proceedings of the DARPA/RL Planning Initiative*, 1994. 2

[16] D. Calisi, A. Farinelli, L. Iocchi, and D. Nardi. Autonomous exploration for search and rescue robots. In *Proceedings of the International Conference on Safety and Security Engineering (SAFE)*, pages 305–314, 2007. DOI: 10.2495/SAFE070301. 88

[17] S. Chakravarty, and Y. Shahar. Specification and detection of periodicity in clinical data. *Methods of Information in Medicine*, 40:410–420, 2001. 92

[18] S. Chien, G. Rabideau, R. Knight, R. Sherwood, B. Engelhardt, D. Mutz, T. Estlin, B. Smith, F. Fisher, T. Barrett, G. Stebbins, and D. Tran. ASPEN - automated planning and scheduling for space mission operations. In *Space Ops*, 2000. 82

[19] M. Clark. *The Gantt Chart, a Working Tool of Management*. Sir Issac Pittman and Sons, London, 1942. 2

[20] D. Colbry, B. Peintner, and M. E. Pollack. Execution monitoring with quantitative temporal dynamic bayesian networks. In M. Ghallab, J. Hertzberg, and P. Traverso, editors, *Proceedings of the Sixth International Conference on Artificial Intelligence Planning Systems (AIPS'02)*, pages 194–203, 2002. 90

[21] T.H. Cormen, C.E. Leiserson, and R.L. Rivest. *Introduction to Algorithms*. MIT Press, Cambridge, MA, 1990. 29, 52

[22] G. Cortellessa, A. Cesta, A. Oddi, and N. Policella. User interaction with an automated solver. the case of a mission planner. *PsychNology Journal*, 2(1):140-162, 2004. 82

[23] E. Cox. Fuzzy fundamentals. *IEEE Spectrum*, 29(10):58–61, 1992. DOI: 10.1109/6.158640. 73

[24] G.B. Dantzig. *Linear Programming and Extensions*. Princeton University Press, 1962. 28

[25] T. Dean and K. Kanazawa. Probabilistic temporal reasoning. In H. E. Shrobe, T. M. Mitchell, and R. G. Smith, editors, *Proceedings of the National Conference on Artificial Intelligence (AAAI-1988)*, pages 524–529. AAAI Press / The MIT Press, 1988. 90

[26] R. Dechter, I. Meiri, and J. Pearl. Temporal constraint networks. *Artificial Intelligence*, 49(1-3):61–95, 1991. DOI: 10.1016/0004-3702(91)90006-6. 28, 30, 32, 33, 40, 42, 65

[27] O. Despouys and F. F. Ingrand. Propice-plan: Toward a unified framework for planning and execution. In S. Biundo and M. Fox, editors, *Recent Advances in AI Planning (Proceedings of the 5th European Conference on Planning, ECP-99)*, volume 1809 of *Lecture Notes in Computer Science*, pages 278–293. Springer, 1999. 85

[28] J. Dix, S. Kraus, and V.S. Subrahmanian. Agents dealing with time and uncertainty. In *Proceedings of the First International Joint Conference on Autonomous Agents and Multiagent Systems: part 2 (AAMAS'02)*, pages 912–919. ACM, 2002. DOI: 10.1145/544862.544953. 10, 89

[29] Ch. Dousson, P. Gaborit, and M. Ghallab. Situation recognition: Representation and algorithms. In *Proceedings of the 13th International Joint Conference on Artificial Intelligence - Volume 1 (IJCAI'93)*, pages 166–172. Morgan Kaufmann Publishers, 1993. 89

[30] C. Dousson and P. Le Maigat. Chronicle recognition improvement using temporal focusing and hierarchization. In *Proceedings of the 20th International Joint Conference on Artificial Intelligence (IJCAI'07)*, pages 324–329. Morgan Kaufmann Publishers, 2007. 89, 90

[31] D. Dubois and H. Prade. What are fuzzy rules and how to use them. *Fuzzy Sets and Systems*, 84:169–185, 1996. DOI: 10.1016/0165-0114(96)00066-8. 73

[32] D. Dubois, A. HadjAli, and H. Prade. Fuzziness and uncertainty in temporal reasoning. *Journal of Universal Computer Science*, 9(9):1168–1194, 2003. DOI: 10.3217/jucs-009-09-1168. 46, 48, 77

[33] R. T. Effinger. *Risk-minimizing Program Execution in Robotic Domains*. PhD thesis, Massachusetts Institute of Technology, 2012. 87, 88

[34] M. Falda, F. Rossi, and K. B. Venable. Dynamic consistency of fuzzy conditional temporal problems. *Journal of Intelligent Manufacturing*, 21(1):75–88, 2010. 75, 77

[35] M. Falda, F. Rossi, and K. B. Venable. Fuzzy conditional temporal problems: Strong and weak consistency. *Engineering Applications of Artificial Intelligence*, 21(5):710–722, 2008. DOI: 10.1016/j.engappai.2008.03.010. 73, 74, 76, 77

[36] M. Falda, F. Rossi, and K. B. Venable. Dynamic consistency of fuzzy conditional temporal problems. *Journal of Intelligent Manufacturing*, 21(1):75–88, 2010. DOI: 10.1007/s10845-008-0170-9. 76, 77

[37] S. S. Fatima, M. Wooldridge, and N. R. Jennings. Multi-issue negotiation under time constraints. In *Proceedings of the First International Joint Conference on Autonomous Agents and Multiagent Systems: part 1, (AAMAS '02)*, pages 143–150, ACM, 2002. DOI: 10.1145/544741.544775. 7

[38] M. Fisher, D. Gabbay, and L. Vila. *Handbook of Temporal Reasoning in Artificial Intelligence (Foundations of Artificial Intelligence)*. Elsevier Science Inc., New York, NY, USA, 2005. 9

[39] M. Franceschet and A. Montanari. A combined approach to temporal logics for time granularity. In *Proceedings of the Second International Workshop on Methods for Modalities (M4M)*, 2001. 5

[40] J. Frank, A. Jonsson, R. Morris, and D. Smith. Planning and scheduling for fleets of earth observing satellites. In *6th Intl. Symposium on AI, Robotics, and Automation in Space (i-SAIRAS'01)*, 2001. 69

[41] J. Frank and A. Jonsson. Constraint- based attribute interval planning. *Journal of Constraints*, 8(4), 2003. DOI: 10.1023/A:1025842019552. 82

[42] A. Galton. Temporal logic. In E. N. Zalta, editor, *The Stanford Encyclopedia of Philosophy*. Fall 2008 edition, 2008. 4

[43] D. Goren-Bar, Y. Shahar, M. Galperin-Aizenberg, D. Boaz, and G. Tahan. KNAVE-II: the definition and implementation of an intelligent tool for visualization and exploration of time-oriented clinical data. In *Proceedings of the working conference on Advanced visual interfaces*, AVI '04, pages 171–174, New York, NY, USA, 2004. ACM. DOI: 10.1145/989863.989889. 91

[44] F. Heintz. Chronicle Recognition in the WITAS UAV Project - A Preliminary Report. In *Swedish AI Society Workshop (SAIS2001)*, 2001. 89

[45] J.W. Herrmann. *Handbook of production scheduling*. International series in operations research & management science. Springer, 2006. 2

[46] L. Hunsberger. Distributing the control of a temporal network among multiple agents. In *Proceedings of the Second International Joint Conference on Autonomous Agents and Multiagent Systems (AAMAS'03)*, pages 899-906. ACM, 2003. DOI: 10.1145/860575.860720. 88

[47] L. Hunsberger. Fixing the semantics for dynamic controllability and providing a more practical characterization of dynamic execution strategies. In *Proceedings of 16th International Symposium on Temporal Representation and Reasoning (TIME)*, pages 155–162. IEEE Computer Society, 2009. DOI: 10.1109/TIME.2009.25. 64, 65

[48] L. Hunsberger. A fast incremental algorithm for managing the execution of dynamically controllable temporal networks. In *Proceedings of 17th International Symposium on Temporal Representation and Reasoning (TIME)*, pages 121–128. IEEE Computer Society, 2010. DOI: 10.1109/TIME.2010.16. 64

[49] M. Iri. *Network flow, transportation, and scheduling; theory and algorithms*. Mathematics in science and engineering. Academic Press, 1969. 2, 3

[50] P. Jonsson and A.A. Krokhin. Complexity classification in qualitative temporal constraint reasoning. *Artificial Intelligence*, 160(1-2):35–51, 2004. DOI: 10.1016/j.artint.2004.05.010. 25

[51] M. D. Johnston and B. J. Clement. Automating deep space network scheduling and conflict resolution. In H. Nakashima, M. P. Wellman, G. Weiss, and P. Stone, editors, *Proceedings of the Fifth International Joint Conference on Autonomous Agents and Multiagent Systems (AAMAS'06*, pages 1483–1489. ACM, 2006. DOI: 10.1145/1160633. 82

[52] A. K. Jónsson, R. A. Morris, and L. Pedersen. Autonomy in space: Current capabilities and future challenge. *AI Magazine*, 28(4):27–42, 2007. DOI: 10.1109/AERO.2007.352852. 85

[53] H. Kautz and P. Ladkin. Integrating metric and qualitative temporal reasoning. In *Proceedings of the National Conference on Artificial Intelligence (AAAI-1991)*, pages 241–246. AAAI Press, 1991. 41

[54] L.G. Khachiyan. A polynomial algorithm in linear programming. *Soviet Mathematics Doklady*, pages 191–194, 1979. 28

[55] L. Khatib, P. H. Morris, R. A. Morris, and F. Rossi. Temporal constraint reasoning with preferences. In B. Nebel, editor, *Proceedings of the Seventeenth International Joint Conference on Artificial Intelligence (IJCAI 2001)*, pages 322–327. Morgan Kaufmann, 2001. 48, 49, 50, 77

[56] L. Khatib, P. H. Morris, R. A. Morris, and K. B. Venable. Tractable Pareto optimization of temporal preferences. In G. Gottlob and T. Walsh, editors, *Proceedings of the Eighteenth International Joint Conference on Artificial Intelligence (IJCAI-03)*, pages 1289–1294. Morgan Kaufmann, 2003. 56, 77

[57] L. Khatib, P. H. Morris, R. Morris, F. Rossi, A. Sperduti, and K. B. Venable. Solving and learning a tractable class of soft temporal constraints: Theoretical and experimental results. *AI Communications*, 20(3):181–209, 2007. 49, 51, 52, 53, 54, 77

[58] P. Kim, B. C. Williams, and M. Abramson. Executing reactive, model-based programs through graph-based temporal planning. In *Proceedings of the 17th International Joint Conference on Artificial Intelligence - Volume 1 (IJCAI'01)*, pages 487–493, Morgan Kaufmann Publishers, 2001. 87

[59] M. Koubarakis. From local to global consistency in temporal constraint networks. *Theoretical Computer Science*, 173(1):89–112, 1997. DOI: 10.1016/S0304-3975(96)00192-2. 25, 40

[60] R. Kowalski and M. Sergot. A logic-based calculus of events. *New Generation Computing*, 4(1):67–95, January 1986. DOI: 10.1007/BF03037383. 4

[61] A.A. Krokhin, P. Jeavons, P. Johsson. Reasoning about temporal relations: The tractactable subalgebras of Allen's interval algebra. *Journal of ACM*, 50(5):591–640, 2003. DOI: 10.1145/876638.876639. 24, 25

[62] A.A. Krokhin, P. Jeavons, P. Johsson. Constraint satisfaction problems on intervals and lengths. *SIAM Journal of Discrete Mathematics*, 17(3):453–477, 2004. DOI: 10.1137/S0895480102410201. 41

[63] P. Laborie. Resource temporal networks: Definition and complexity. In *Proceedings of the 18th International Joint Conference on Artificial Intelligence (IJCAI'03)*, pages 948–953. Morgan Kaufmann, 2003. 42

[64] P.B. Ladkin and R.D. Madux. On binary constraint networks. Technical report, Kerstel Institute, Palo Alto, 1989. 16

[65] C. C. Lee. Fuzzy logic in control systems: fuzzy controllers. *IEEE Trans. on Sys., Man and Cybern.*, 2092:404–435, 1990. DOI: 10.1109/21.52552. 73

[66] G. Ligozat. On generalized interval calculi. In T. L. Dean and K. McKeown, editors, *Proceedings of the National Conference on Artificial Intelligence (AAAI-1991)*, pages 234–240. AAAI Press / The MIT Press, 1991. 92

[67] X. Ling and D. S. Weld. Temporal information extraction. In M. Fox and D. Poole, editors, *Proceedings of the National Conference on Artificial Intelligence (AAAI-2010)*. AAAI Press, 2010. 92

[68] A.K. Mackworth. Consistency in networks of relations. *Artificial Intelligence*, 8:99–118, 1977. DOI: 10.1016/0004-3702(77)90007-8. 34

[69] A.K. Mackworth. Constraint satisfaction. In S. C. Shapiro, editor, *Encyclopedia of AI (second edition)*, volume 1, pages 285–293. John Wiley & Sons, 1992. 49

[70] J. Malik and T. O. Binford. Reasoning in time and space. In *International Joint Conference on Artificial Intelligence (IJCAI'83)*, pages 343–345. Morgan Kaufmann Publishers, 1983. 39

[71] M. Maniadakis, P. Trahanias, and J. Tani. Explorations on artificial time perception. *Neural Networks*, 22:509–517, 2009. DOI: 10.1016/j.neunet.2009.06.045. 7

[72] C. McGann, F. Py, K. Rajan, J. P. Ryan, and R. Henthorn. Adaptive control for autonomous underwater vehicles. In D. Fox and C. P. Gomes, editors, *Proceedings of the National Conference on Artificial Intelligence (AAAI-1988)*, pages 1319–1324. AAAI Press, 2008. 88

[73] I. Meiri. Combining qualitative and quantitative constraints in temporal reasoning. *Artificial Intelligence*, 87:343–385, 1996. DOI: 10.1016/0004-3702(95)00109-3. 24, 41

[74] M.D. Moffitt, B. Peintner, M.E. Pollack. Augmenting disjunctive temporal problems with finite-domain constraints. In *Proceedings of the 20th National Conference on Artificial Intelligence (AAAI-2005)*, pages 1187–1192. AAAI Press, 2005. 42

[75] P. Morris, R. Morris, L. Khatib, S. Ramakrishnan, and A. Bachmann. Strategies for global optimization of temporal preferences. In M. Wallace, editor, *Proceeding of the 10th International Conference on Principles and Practice of Constraint Programming (CP-04)*, volume 3258 of *Lecture Notes in Computer Science*, pages 588–603. Springer, 2004. 56, 57, 77

[76] P. H. Morris, N. Muscettola, and T. Vidal. Dynamic control of plans with temporal uncertainty. In B. Nebel, editor, *17th International Joint Conference on Artificial Intelligence (IJCAI 2001)*, pages 494–502. Morgan Kaufmann, 2001. 61, 62, 63, 64, 72, 77

[77] P. Morris. A structural characterization of temporal dynamic controllability. In *Principles and Practice of Constraint Programming - CP 2006*, volume 4204 of *Lecture Notes in Computer Science*, pages 375–389. Springer, 2006. DOI: 10.1007/11889205_28. 64, 68

[78] P. H. Morris and N. Muscettola. Managing temporal uncertainty through waypoint controllability. In T. Dean, editor, *Proceedings of the 16th International Joint Conference on Artificial Intelligence (IJCAI'99)*, pages 1253–1258. Morgan Kaufmann, 1999. 61, 77

[79] P. H. Morris and N. Muscettola. Temporal dynamic controllability revisited. In *20th National Conference on Artificial Intelligence (AAAI 2005)*, pages 1193–1198. AAAI Press / The MIT Press, 2005. 61, 63, 64, 77

[80] R. A. Morris and L. Khatib. Constraint reasoning about repeating events: Satisfaction and optimization. *Computational Intelligence*, 16(2):257–278, 2000. DOI: 10.1111/0824-7935.00113. 92

[81] N. Muscettola, P. Morris, B. Pell, and B. Smith. Issues in temporal reasoning for autonomous control systems. In *Proceedings of the Second International Conference on Autonomous Agents (AGENTS '98)*, pages 362–368. ACM Press, 1997. DOI: 10.1145/280765.280862. 7, 86

[82] N. Muscettola, P. Pandurang Nayak, B. Pell, and B. C. Williams. Remote agent: To boldly go where no AI system has gone before, 1998. *Artificial Intelligence*, 103(1-2): 5–47, 1998. DOI: 10.1016/S0004-3702(98)00068-X. 85

[83] N. Muscettola, B. Smith, Ch. Fry, S. Chien, K. Rajan, G. Rabideau, and D. Yan. On-board planning for New Millennium Deep Space One autonomy. In *Proceedings of Aerospace Conference*, pages 303-318, IEEE, 1997. DOI: 10.1109/AERO.1997.574421. 82

[84] G. Nagypál and B. Motik. A fuzzy model for representing uncertain, subjective, and vague temporal knowledge in ontologies. In *CoopIS/DOA/ODBASE*, volume 2888 of *Lecture Notes in Computer Science*, pages 906–923. Springer, 2003. DOI: 10.1007/978-3-540-39964-3_57. 46, 48, 77

[85] B. Nebel and H.-J. Bürckert. Reasoning about temporal relations: A maximal tractable subclass of Allen's interval algebra. *Journal of the ACM*, 42(1):43–66, 1995. DOI: 10.1145/200836.200848. 24, 25, 47

[86] H. J. Ohlbach. Relations between fuzzy time intervals. In *Proceedings of 11th International Symposium on Temporal Representation and Reasoning (TIME 2004)*, pages 44–51. IEEE Computer Society, 2004. DOI: 10.1109/TIME.2004.1314418. 46, 48, 77

[87] J. Pearl R. Dechter, I. Meiri. Temporal constraint networks. *Artificial Intelligence*, 49(1-3):61–95, 1991. DOI: 10.1016/0004-3702(91)90006-6. 39

[88] B. Peintner and M. E. Pollack. Low-cost addition of preferences to DTPs and TCSPs. In *Proceedings of the 19th National Conference on Artificial Intelligence (AAAI'04)*, pages 723–728. AAAI Press / The MIT Press, 2004. 55

[89] B. Peintner and M. E. Pollack. Anytime, complete algorithm for finding utilitarian optimal solutions to STPPs. In *Proceedings of the 20th National Conference on Artificial Intelligence (AAAI 2005)*, pages 443–448. AAAI Press / The MIT Press, 2005. 57, 77

[90] L. R. Planken. New algorithms for the simple temporal problem. *Master Thesis.* Delft University of Technology, 2008. 39

[91] M. E. Pollack, C. E. McCarthy, I. Tsamardinos, S. Ramakrishnan, L. Brown, S. Carrion, D. Colbry, Ch. Orosz, and B. Peintner. Autominder: A planning, monitoring, and reminding assistive agent, In *Proceeding of the International Conference on Intelligent Autonomous Systems,* 2002. 90

[92] R. Posenato, L. Hunsberger and C. Combi. The dynamic controllability of conditional STNS with uncertainty. In *Proceedings of ICAPS 2012 Workshop on The Planning and Plan Execution for Real-World Systems: Principles and Practices (PlanEx),* pages 121–128, 2012. 67

[93] R. Posenato, L. Hunsberger and C. Combi. An algorithm for checking the dynamic controllability of a conditional simple temporal network with uncertainty. In *Proceedings of the 5th International Conference on Agents and Artificial Intelligence (ICAART-2013),* pages 144-156, 2013. 68

[94] A. Prior. *Past, Present and Future.* Oxford: Clarendon Press, 1967. DOI: 10.1093/acprof:oso/9780198243113.001.0001. 3

[95] J. Pustejovsky, J. Castaño, R. Ingria, R. Saurí, R. Gaizauskas, A. Setzer, and G. Katz. TimeML: Robust Specification of Event and Temporal Expressions in Text, In *Proceedings of the Fifth International Workshop on Computational Semantics (IWCS-5),* pages 1–11, 2003. 92

[96] D. Rosenberg and A. Grafton. *Cartographies of Time: A History of the Timeline.* Princeton Architectural Press, 2010. 1

[97] F. Rossi, K. B. Venable, and N. Yorke-Smith. Uncertainty in soft temporal constraint problems: a general framework and controllability algorithms for the fuzzy case. *Journal of AI Research,* 27:617–674, 2006. DOI: 10.1613/jair.2135. 69, 70, 71, 72, 77

[98] S. J. Russell and P. Norvig. *Artificial intelligence: a modern approach.* Prentice-Hall, Inc., Upper Saddle River, NJ, USA, 1996. 6

[99] E. Sacerdoti. Planning in a hierarchy of abstraction spaces. *Artificial Intelligence,* 5(2):115–135, 1974. DOI: 10.1016/0004-3702(74)90026-5. 39

[100] T. Schiex. Possibilistic Constraint Satisfaction problems or "How to handle soft constraints?". In D. Dubois and M. P. Wellman, editors, *8th Annual Conference on Uncertainty in Artificial Intelligence (UAI'92),* pages 268–275. Morgan Kaufmann, 1992. 49

[101] G. Schindler and F. Dellaert. Probabilistic temporal inference on reconstructed 3d scenes. In *Proceedings of the Conference on Computer Vision and Pattern Recognition (CVPR)*, pages 1410–1417. IEEE, 2010. DOI: 10.1109/CVPR.2010.5539803. 91

[102] E. Schwalb and R. Dechter. Processing disjunctions in temporal constraint networks. *Artificial Intelligence*, 93(1-2):29–61, 1997. DOI: 10.1016/S0004-3702(97)00009-X. 33, 34, 35, 40

[103] Y. Shahar. A framework for knowledge-based temporal abstraction. *Artificial Intelligence*, 90(1-2):79–133, 1997. DOI: 10.1016/S0004-3702(96)00025-2. 6

[104] R. Sherwood, A. Govindjee, D. Yan, G. Rabideau, S. Chien, and A. Fukunaga. ASPEN: EO-1 mission activity planning made easy. Technical Report, NASA JPL, 1997. 82

[105] R. Shostak. Deciding linear inequalities by computing loop residues. *Journal of the ACM*, 28(4):769–779, 1981. DOI: 10.1145/322276.322288. 29

[106] H. Simon. Two heads are better than one: the collaboration between AI and OR. *Interfaces*, 17:8–15, 1987. DOI: 10.1287/inte.17.4.8. 4

[107] K. Stergiou and M. Koubarakis. Backtracking algorithms for disjunctions of temporal constraints. *Artificial Intelligence*, 120(1):81–117, 2000. DOI: 10.1016/S0004-3702(00)00019-9. 36, 37, 39, 42

[108] D. Strömberg. Decision-making using temporal reasoning. In *IJCAI-99 Workshop on Teams*, 1999. 89

[109] M. Tatu and M. Srikanth. Experiments with reasoning for temporal relations between events. In *Proceedings of the 22nd International Conference on Computational Linguistics - Volume 1, (COLING '08)*, pages 857–864. Association for Computational Linguistics, 2008 92

[110] I. Tsamardinos. A probabilistic approach to robust execution of temporal plans with uncertainty. In I. P. Vlahavas and C. D. Spyropoulos, editors, *Methods and Applications of Artificial Intelligence, Second Hellenic Conference on AI (SETN 2002)*, volume 2308 of *Lecture Notes in Computer Science*, pages 97–108. Springer, 2002. 59

[111] I. Tsamardinos, T. Vidal, and M. E. Pollack. CTP: A new constraint-based formalism for conditional, temporal planning. *Constraints*, 8(4):365–388, 2003. DOI: 10.1023/A:1025894003623. 65, 66, 67, 76, 77

[112] I. Tsamardinos and N. Muscettola. Fast transformation of temporal plans for efficient execution. In *Proceedings of the Thirteenth National Conference on Artificial Intelligence (AAAI-1998)*, pages 254–261, 1998. 88

[113] P. van Beek. Approximation algorithms for temporal reasoning. In *Proceedings of the 11th International Joint Conference on Artificial Intelligence (IJCAI'89) - Volume 2*, pages 1291–1296. Morgan Kaufmann, 1989. 16, 25

[114] P. van Beek. Reasoning about qualitative temporal information. *Artificial Intelligence*, 58(1-2):297–326, 1992. DOI: 10.1016/0004-3702(92)90011-L. 16

[115] T. Vidal and H. Fargier. Handling contingency in temporal constraint networks: from consistency to controllabilities. *Journal of Experimental and Theoretical Artificial Intelligence*, 11(1):23–45, 1999. DOI: 10.1080/095281399146607. 57, 58, 59, 60, 61, 77

[116] M. Vilain and H. Kautz. Constraint propagation algorithms for temporal reasoning. In *National Conference on Artificial Intelligence (AAAI'86)*, pages 377–382. AAAI Press, 1986. 15, 17, 20, 23, 24, 25, 41

[117] M. Vilain, H. Kautz, P. van Beek. Constraint propagation algorithms for temporal reasoning: A revised report. In J. de Kleer D. Weld, editors, *Readings in Qualitative Reasoning about Physical Systems*, pages 373–381. Morgan Kaufmann, 1989. 25

[118] Wikipedia. Time, 2004. [Online; accessed 22-July-2004]. 1

[119] L. A. Zadeh. Calculus of fuzzy restrictions. *Fuzzy Sets and their Applications to Cognitive and Decision Processes*, pages 1–40, 1975. 54

Authors' Biographies

ROMAN BARTÁK

Roman Barták is a professor at Charles University, Prague (Czech Republic). He leads the Constraint Satisfaction and Optimization Research Group that performs basic and applied research in the areas of satisfiability and discrete optimization problems. His work focuses on techniques of constraint satisfaction and their application to planning and scheduling. The research results are used in products of ILOG, Visopt, and ManOPT/Entellexi. Prof. Barták is teaching courses on artificial intelligence, planning, scheduling, and constraint programming at Charles University and he presented several tutorials on these topics at major conferences such as IJCAI, AAAI, ICAPS, SAC etc.; he is author of the On-line Guide to Constraint Programming (#2 source for Constraint Programming in Google).

ROBERT A. MORRIS

Robert A. Morris is a senior researcher in Computer Science in the Exploration Technology Directorate, Intelligent Systems Division at NASA Ames Research Center. His primary professional goal is the application of advanced AI technology in planning, scheduling, and plan execution to the next generation of NASA's exploration systems. His primary research interests include temporal constraint-based reasoning for automated planning and scheduling.

K. BRENT VENABLE

K. Brent Venable is an associate professor in the Department of Computer Science of Tulane University and research scientist at IHMC, the Florida Institute of Human and Machine Cognition. In the past she has been an assistant professor in the Department of Pure and Applied Mathematics at the University of Padova (Italy). Her main research interests are within artificial intelligence and regard, in particular, compact preference representation formalisms, computational social choice, temporal reasoning and, more in general, constraint-based optimization. Her list of publications includes more than 70 papers including journals and proceedings of the main international conferences on the topics relevant to her interests. She is involved in a lively international scientific exchange and, among others, she collaborates with researchers from NASA Ames, SRI International, NICTA-UNSW (Australia), University of Amsterdam (The Nederlands), 4C (Ireland), and Ben-Gurion University (Israel).

Printed in the United States
by Baker & Taylor Publisher Services